Christmas
from Howard &
Thelma
Xmas 2000

LAND ROVER

CONVERSIONS & APPLICATIONS SINCE 1948

The world's most versatile vehicle at work

Richard de Roos

LAND ROVER

CONVERSIONS & APPLICATIONS SINCE 1948

The world's most versatile vehicle at work

Richard de Roos

MOTOR RACING PUBLICATIONS LTD
Unit 6, The Pilton Estate, 46 Pitlake, Croydon CRO 3RY, England

First published 1995

British Library Cataloguing in Publication Data
Roos, Richard de
Land Rover: Conversions and Applications
Since 1948
I. Title
629.2222

ISBN 0-947981-87-X

Typesetting and origination by Keele University Press, Staffordshire
Printed in Great Britain by The Amadeus Press Ltd, Huddersfield, West Yorkshire

Cover illustrations:
Front: Tipper conversion by Special Vehicles based on the Defender 130 Crew Cab (photo: P D Stevens & Sons Ltd); Land Rover V8 6x6 fire tender conversion by Carmichael Holdings Ltd (photo: Nick Dimbleby); fully air portable and demountable military ambulance unit by Shanning POD Ltd (photo: Shanning POD Ltd). Rear: Camper unit with accommodation for four people mounted on a Defender 130 Crew Cab Pick Up (photo: Autarkia Camperunits); reinforced polyester scraper blade designed to fit the Land Rover One Ten (photo: Nido Universal Machines bv).

Contents

Introduction

The Land Rover is the world's most versatile vehicle and one of the motor industry's biggest success stories. Originally designed with multi-purpose intentions, based on the example of the US Army Jeep, the resulting vehicle appeared in 1947. No-one then could possibly have envisaged that nearly 50 years later the Land Rover's popularity amongst its extremely diverse customers would remain undiminished.

This book is a pictorial celebration of the Land Rover/ Defender's extraordinary versatility, portraying the many conversions and applications possible plus concomitant special equipment manufactured. Divided into two sections, the first depicts the production models, for which a comprehensive accessory set is optionally available comprising either towing bracket, power take-off, winch or roof-rack, to adapt the vehicle to the job. Yet these items do not in themselves constitute 'conversions'.

Conversions are instead effected by radical modifications – a role performed by, amongst others, the Special Vehicles Department within Land Rover Limited, who are responsible for the original prototype engineering, development, manufacture and improvement of specially adapted Land Rovers.

Section Two is devoted to such resulting conversions and their applications, featuring virtually everything from the Land Rover's familiar guises in the military and emergency services to its lesser-known road/rail, amphibious-vehicle, golfball-flight-demonstrator, cushion-craft-vehicle and X-ray unit conversions – to name but a few! Each of these, plus many more, are arranged under their relevant, alphabetically-listed, application headings where they are then described according to their particular task. The year for each photograph represents that in which it was taken and when the conversion was in production, many for the first time; in some cases, however, this denotes the only year it was in production. The year beside an author's sketch represents that in which it was completed.

The number of Land Rover/Defender conversion and application possibilities is continually expanding and therefore defies definitive collation. However, it is very much hoped that the following depictions of this vehicle's many facets will stimulate the reader's interest in, and endorse recognition of, this foremost 4x4.

January 1995 Richard de Roos

Acknowledgements

This book would never have been possible without the invaluable help and assistance provided by the following individuals and organizations.

Particular thanks must be extended to Nick Dimbleby, who supplied a large number of photographs from his private collection. For all their support and encouragement in my research I also wish to thank those institutions and companies involved in conversion and special equipment activities, as well as the following who provided drawings, photographs and information:

Abbey Electronics Ltd, Action Mobil of Austria, Afdeling Maritieme Historie van de Marinestaf, Agri-Visual Ltd, Allam Generators Ltd, E Allman & Company Ltd, G Hydes at Amphibious Trials and Training Unit Royal Marines (ATTURM), Angloco Ltd, H C B Angus Ltd, Autarkia Camperunits, Dutch franchised dealer Automobielcentrale 'de Uiver' bv, The Automobile Association, Avon Inflatables Ltd, Binks-Bullows Ltd, BOC Ltd, Brimec (UK) Ltd, Bristol & West Photography, The British Drilling Association Ltd, British Motor Industry Heritage Museum/Rover Group, Caranex, Carmichael Ltd, Cars Gravemeijer (Rover Nederland bv), Chubb Group Services Ltd, Clark Masts Teksam Ltd, Commando Tactische Luchtstrijdkrachten, Compair Holman, Courtaulds Aerospace Ltd, J A Cuthbertson Ltd, R Dekker, Dennis Eagle Ltd, Dixon Bate Ltd, Dormobile Ltd, Econ Group Ltd, E V Engineering Ltd, F K I Bradbury Ltd, Forestry Commission, Fotografie Lex Klimbie, Francome Fabrications Ltd, Froude Consine, Glover Webb Ltd, Hands-England Division of James Howden & Co Ltd, Harry Andriessen (Don van der Vaart Fotoproductions bv), Hayters plc, Headquarters Dutch Marines, Heathrow Airport Ltd, Hubbard Transport Refrigeration Ltd, Hymatic Engineering Company Ltd, Ifor Williams Trailers Ltd, International Transport & Management, Howard Jones, Kelvin Hughes Ltd, Marcus Kelly, King Trailers Ltd, Korps Rijkspolitie Dienst Luchtvaart, Kronenburg bv, Land Rover Ltd, Land Rover Australia, Land Rover of North America Inc, Macclesfield Motor Bodies (UK) Ltd, MacNeillie & Son Ltd, Marshall of Cambridge (Engineering) Ltd, *Materialist-Magazine*, Mechanische Werkplaats Kemp, Merseyside Passenger Transport Authority – Mersey Tunnels, Ministerie van Defensie Directie Voorlichting, F W McConnel Ltd, H Molter, National Motor Museum Beaulieu, Nido Universal Machines bv, Normalair-Garrett Ltd, Pegson Ltd, Penfold Golf Ltd, Penman Engineering Ltd, Cliff Petts, Photographic Department Leeuwarden Air Force Base, Pilcher-Greene Ltd, Powered Access Ltd, Premier Hazard Systems Ltd, Pye TVT Ltd, R F D Ltd, Raydel Engineering Ltd, Reynolds Boughton Ltd, E van Rhee (Sieberg bv), Rover Nederland bv, Sambell Engineering Ltd, Scottorn Trailers Ltd, Selectokil Ltd, Shanning POD Ltd, Short Brothers plc, Simon Dudley, Simon Gloster Saro Ltd, Sisis Equipment Ltd, S M C Engineering (Bristol) Ltd, A Smith Gt Bentley Ltd, P D Stevens & Sons Ltd, Tank Museum Wareham, James Taylor, Tecalemit Garage Equipment Co Ltd, Terberg Specials bv, Tooley Electro Mechanical Co Ltd, Transport Resource Management Ltd, Truckman Ltd, H Uitdenbogerd, V F Specialist Vehicles Ltd, Vickers plc, Wadham Stringer (Coachbuilders) Ltd, Wickham Rail Ltd, Will Engineer Ltd, Zumro bv, Zumro (UK) Ltd.

R d R

Section One:

LAND ROVER Production models

Series I (1948–1958)

Land Rover 86in.
(Afdeling Maritieme Historie van de Marinestaf Den Haag)

(Note: All silhouetted drawings reproduced in Section One are from Land Rover Ltd)

Land Rover 107in.
(Nick Dimbleby)

Series II and IIA (1958–1971)

Land Rover 109in FC.
(Nick Dimbleby)

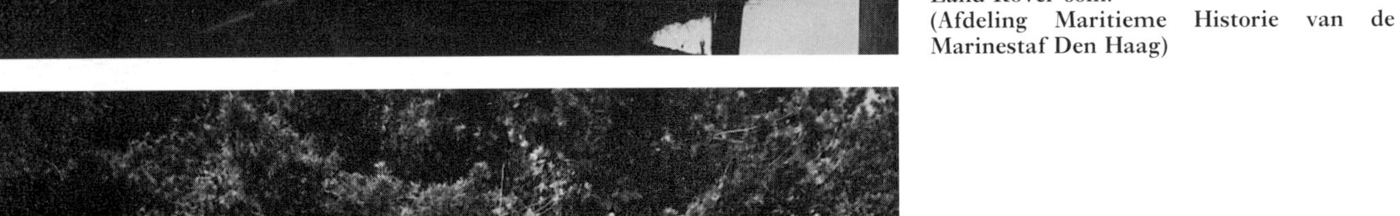

Land Rover 88in.
(Afdeling Maritieme Historie van de Marinestaf Den Haag)

Land Rover 109in.
(R de Roos)

These silhouettes, depicting Land Rovers with short and long wheelbases, normal and forward control driving positions, bare chassis with or without cabs and an almost bewildering choice of body styles ranging from heavy-duty trucks and vans to estate car-type vehicles with a variety of door configurations, were used as part of the company's publicity material in 1965. (Land Rover Ltd)

Series III (1971-1983)

Land Rover 88in. (Land Rover Ltd)

Land Rover 88in Lightweight. (R de Roos)

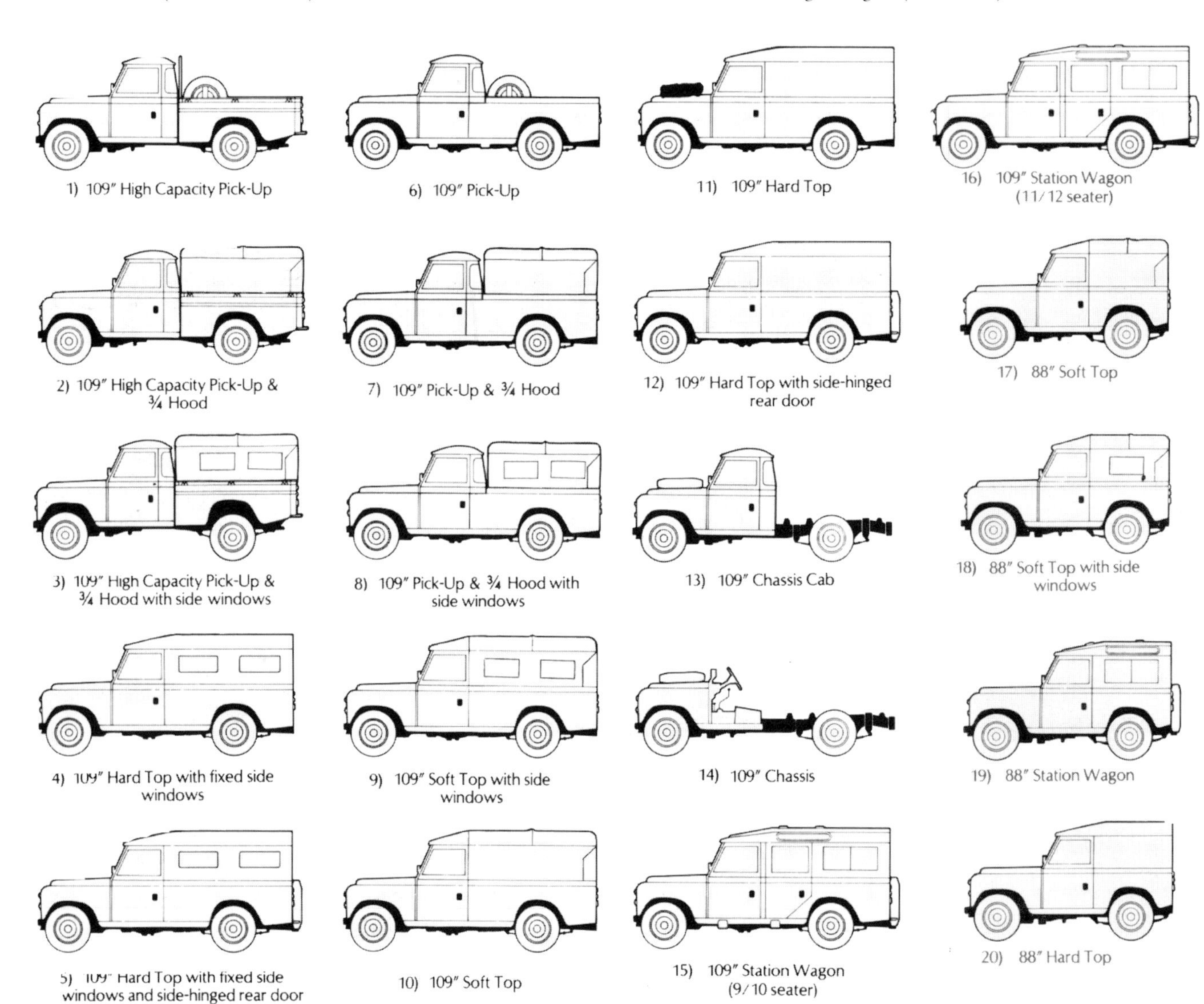

Land Rover 101in Forward Control 1 Ton with Marshall of Cambridge ambulance body. (Nick Dimbleby)

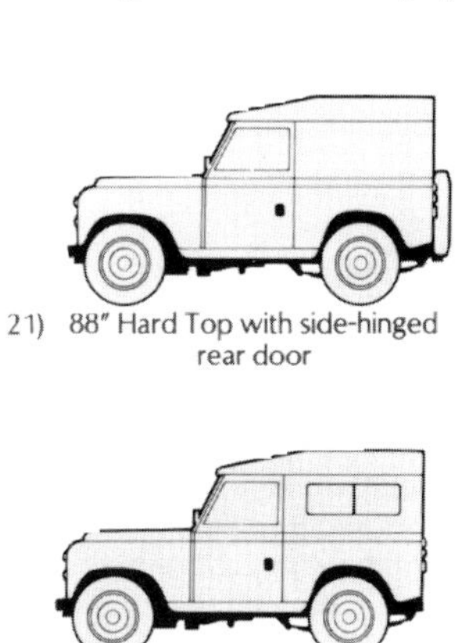

21) 88″ Hard Top with side-hinged rear door

26) 88″ Chassis

22) 88″ Hard Top with sliding side windows

27) 88″ Chassis Cab

23) 88″ Pick-Up & ¾ Hood and side windows

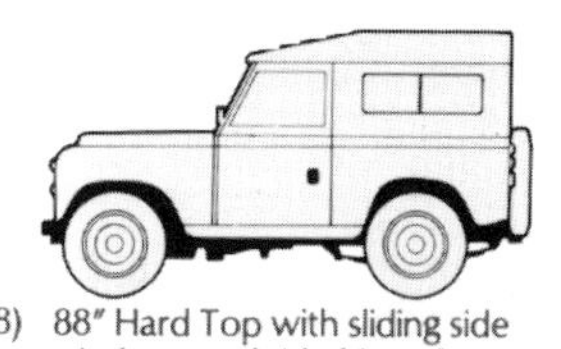

28) 88″ Hard Top with sliding side windows and side-hinged rear door

24) 88″ Pick-Up

29) 88″ Hard Top with fixed side windows

25) 88″ Pick-Up & ¾ Hood

30) 88″ Hard Top with fixed side windows and side-hinged rear door

Land Rover 109in. (Fotografie Lex Klimbie)

Land Rover 109in V8. (R de Roos)

Ninety and One Ten (1983-on)

Land Rover One Ten with factory-mounted safari cage by Safety Devices. (Land Rover of North America Inc)

The extreme versatility of the Land Rover and the variety of factory-built bodies available on its different-wheelbase chassis is clearly illustrated by the profile drawings from the company's publicity material which have been reproduced on these and the previous three pages. But even the choice of eight alternative bodies shown on the Ninety platform on the next page and the 13 on the One Ten depicted here are only a starting point for many specialist users, and the six outline drawings below offer just a small taste of the many conversions and applications which are the subject of the greater part of this book. (Land Rover Ltd)

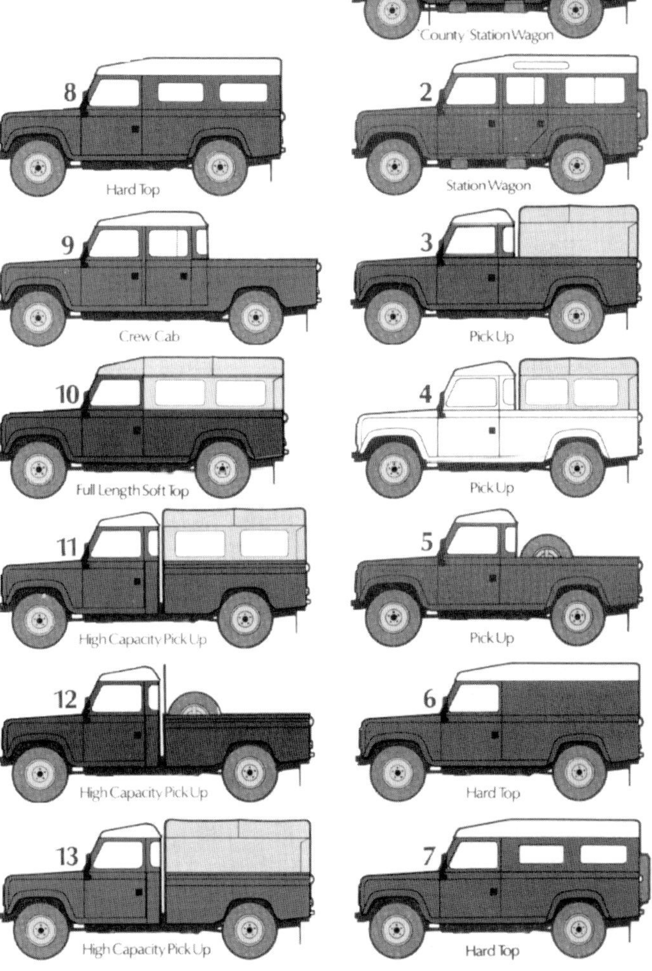

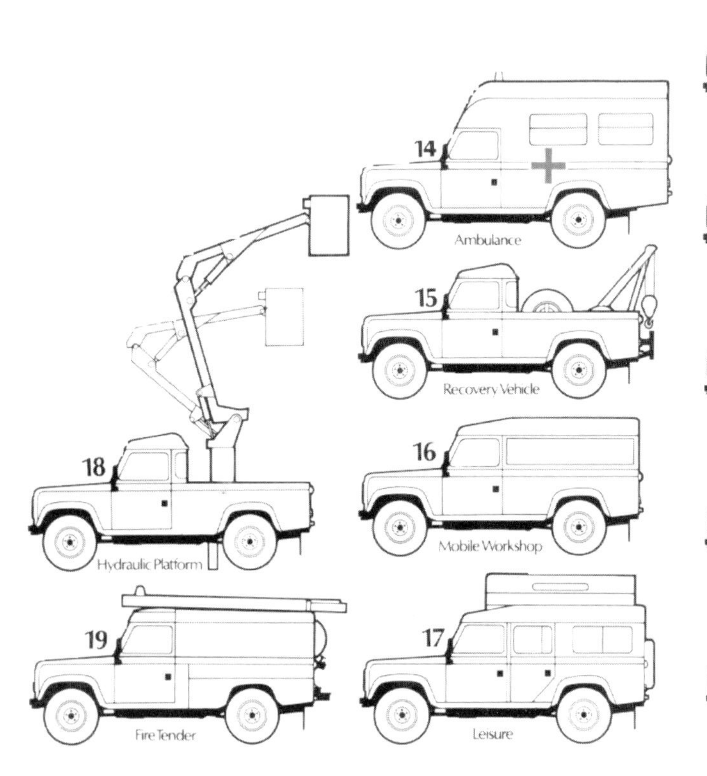

Land Rover One Ten Hard Top. (Land Rover Ltd)

Land Rover Ninety. (Land Rover Ltd)

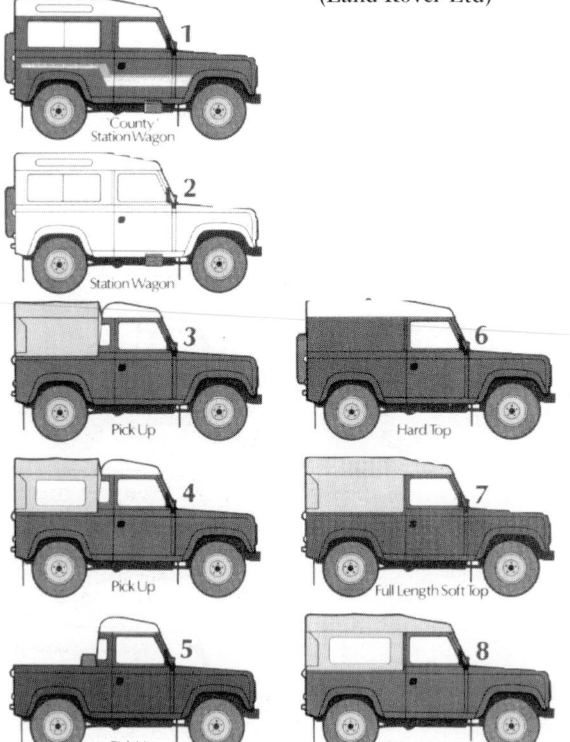

Land Rover/Defender Ninety. (Land Rover Ltd)

Land Rover Ninety Pick Up. (Land Rover Ltd)

Land Rover One Ten Station Wagon. (Land Rover Ltd)

Range Rover (1970-on)

Classic Range Rover.
(R de Roos)

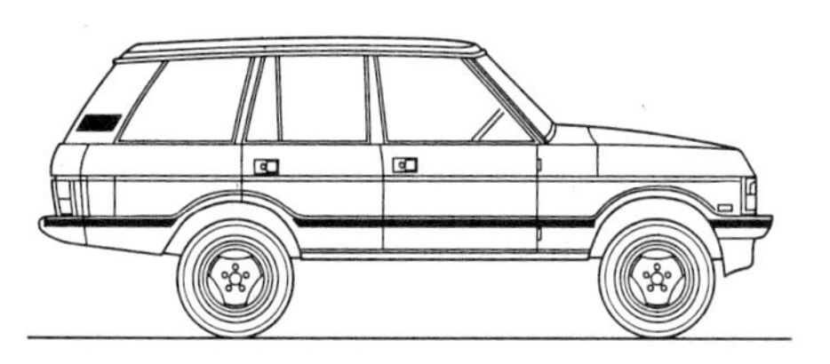

* Classic Range Rover

Classic Range Rover LSE.
(Nick Dimbleby)

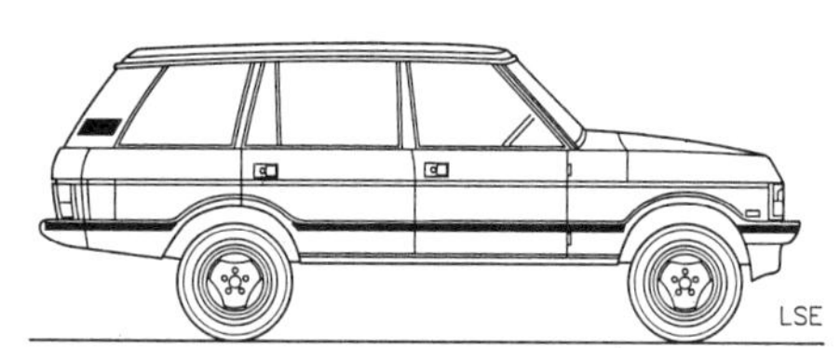

* Classic Range Rover LSE

Range Rover 4.0 V8 SE.
(Land Rover Ltd)

Discovery (1989-on)

Land Rover/Discovery three-door.
(Land Rover Ltd)

Land Rover/Discovery five-door.
(Nick Dimbleby)

* Land Rover/Discovery

Land Rover/Discovery five-door ES.
(Land Rover Ltd)

2862
UE

Section Two:

LAND ROVER
Conversions and Applications

1

1968 shot of the company's Solihull site during one of the first exhibitions of Land Rover applications. From left to right the Series IIA conversions are as follows: Back row – Simon hydraulic platform, Dixon Bate articulated workshop-trailer, Edghill conveyor. Third row – Cintec mobile cinema, Short armoured patrol, Forward Control Consolidated Pneumatics compressor. Second row – MAP Winch, Broom Wade compressor, emergency ambulance, Herbert Lomas ambulance, Dormobile motor caravan, Forward Control Angus fire engine. Front row – McConnel sawbench trailer, Atkinson snow blade, Evers & Wall crop-sprayer, Harvey Frost vehicle recovery crane, Chubb fire engine, Normalair refrigerator.
(1: The National Motor Museum, Beaulieu, 1968)

Agricultural work...

2

3

The Land Rover – as its name implies – was originally intended for agricultural application, and even before the project commenced in 1946, The Rover Company Ltd had decided that in designing a vehicle for the farmer it would have to have four-wheel drive, provision for power take-off and be able to perform many jobs akin to those of the tractor as well as developing sufficient speed for normal road use. These first Land Rovers were installed with many different types of agricultural machinery and equipment including two and three-furrow ploughs, crop-spraying equipment, three-point linkages, pto-driven sawbenches, harrows, threshing machines, hay-makers, elevators, arc welders, agricultural trailers, chain harrows, cultivators, hay-sweepers, binders, combine harvesters, capstan winches, chaff-cutters and milking machines. Crop-spraying equipment for the Land Rover was used for the first time in 1948, manufactured by The Dorman Sprayer Company Ltd. This particular conversion has a 220-litre tank, a pto-driven spray-pump and a 6.8m spray-boom. The market for Land Rover-based crop-spraying equipment was later taken over by Evers & Wall Ltd and was mainly used on Series IIA, IIB Forward Control and III Land Rovers. Such equipment consisted of a hot-dipped galvanized mono-construction tank/frame assembly with two sideglasses and complete drain feature, a pump, hose reels and spray-booms. Two types of spray-boom were available: the 7.3m wide front-mounted type with Hardy plastic fan jets at 0.45m intervals; and the 5.7m front-mounted boom with jets at 0.225m intervals. Each type was fitted with 20mm pipes, which allowed excellent fluid distribution. This sprayer system was fitted with a rear pto-driven diaphragm pump, which produced 85 litres per minute and had the ability to run dry without damage. Its working pressure was adjustable within the 1–15 bar range through a special hand control, and the recommended speed for the spraying operation with the Land Rover was 10km/h. A 450 or 720-litre tank was suitable for the long-wheelbase Land Rover only. These spraying systems were manufactured with a built-in triple filtration fitment as standard. Special options included a twin-jet assembly with protective shield to reduce drift, and a cantilever 2m boom on parallel linkage for spraying banks and road verges. The conversion is also suitable for the chemical de-icing of airfield runways.
(2, 3: Land Rover Ltd, 1980)

4

Selectokil Ltd has for some time offered an 800-litre Land Rover crop-sprayer for the spreading of weedkiller, insecticide and other types of compatible chemical. The Special Vehicle Department has actually offered this conversion – known as the Land Rover Chemi-Sprayer – as an option. The equipment can be easily and quickly dismantled from the vehicle to enable it to return to normal use. All the controls and pipes carrying the spray liquid are kept entirely outside the cab, all boom-spraying functions being operated from a switch box powered from the vehicle's electrical supply. Hand-lance work is carried out using the external controls on the rear of the sprayer's tank unit.
(4: Selectokil Ltd, 1990)

5

6

7

8

Undoubtedly, one of the strangest and most interesting Land Rover conversions ever built was the Series II 109in Pick Up-based cushion-craft crop-sprayer by Vickers-Armstrong in 1962. Two big fans, driven by a second engine, created an air cushion under the vehicle so that the wheel pressure on the ground was adjustable according to the ground's condition. This principle enabled the Land Rover to do crop-spraying work on waterlogged and marshy terrain without sinking. The vehicle was steered by the front wheels – unlike the hovercraft it had neither steering flaps nor a propeller to move forward – and traction was provided by the normal engine and four-wheel drive. The crop-spraying equipment used on this vehicle was manufactured by Evers & Wall Ltd. Although a very expensive experiment, such a conversion was an impractical solution for reducing the Land Rover's ground-pressure and, consequently, it never went into series production. The Vickers-Armstrong factory which manufactured this 'Hover Rover' was the location from where Spitfire fighter aircraft were built during the Second World War. At the time of writing the site is owned by Honda.
(5, 6, 7, 8, 9: Vickers plc, 1962)

9

10

In 1951, E Allman & Company Ltd (Farm Equipment) produced pto-driven spraying equipment known as the Allman Plantector High/Low Volume Sprayer, which could be fitted to the Series I Land Rover. The welded 40-gallon tank was made of heavy-gauge galvanized steel sheet and fitted with two spraying booms whose total spraying width was 5.5m. If these hit an obstruction they were able to fold back and return automatically to the spraying position. The spray-pump was a heavy-duty all-bronze gear pump, fitted with special gears on stainless-steel shafts and capable of developing a normal working pressure of 30psi. For weed control the Land Rover spraying speed was 4mph, a hand-lance being available where access was difficult. The same company also produced rear-power take-off, belt-driven, portable 24in diameter sawbenches, which could be mounted to the Series I Land Rover. In the early Sixties the Allman Plantector Model 60 Field Sprayer was introduced which could be fitted to the Series II Land Rovers. This apparatus was equipped with a 60-gallon capacity galvanized tank mounted in a pressed-steel frame, a Genimec standard patented fingertip control unit and an R V Rollervane spray-pump, direct-driven from the Land Rover centre pto. The adjustable height range for the Model 60's spray boom was from 0.38 to 1.17m. Hand-lances were also available.
(10, 11: E Allman & Company Ltd, 1951)

11

12

To avoid excessive stress in the sidemembers of the Land Rover's chassisframe when a heavy single-axle trailer was attached, Dixon Bate Ltd developed an articulated trailer with a high swan-neck attachment in 1960. Front-wheel traction was not greatly affected as a result. Later, the design for this trailer type was sold to Scottorn Trailers Ltd, who were themselves, in due course, bought out by Reynolds Boughton Ltd.
(12: Land Rover Ltd, 1960)

An articulated semi-trailer manufactured by Reynolds Boughton Ltd attached to a Series III Land Rover Chassis Cab.
(13: Reynolds Boughton Ltd, 1978)

13

14

The rear pto and all-wheel drive made the Land Rover – especially the short-wheelbase type – particularly useful for lawn and sports-field mowing work. Shown here is a trailered, five-unit-powered sports field roller mower manufactured by Sisis Equipment Ltd. All cylinders were driven from the Land Rover engine via pto drive, providing more than sufficient power for the task, and wheel-slip was eliminated even when the ground was soft and wet. Power-driven cylinders also allowed the unit to deal with a wider variety of conditions: in the case of golf course work the fairway could be cut to the required finish; then, when adjusted, the same machine could operate in the semi-rough, cutting grass to the appropriate height, and finally in the rough to top-off taller grass. Transport between sites was simple, as was the attaching and detaching of the mower. The main chassis wheels were of such diameter that it was possible to take the complete unit on the road without incurring the additional cost of a gang-carrier unit or trailer.

15 (14, 15: Sisis Equipment Ltd/Land Rover Ltd, 1976)

The Exhaust Nozzle Sprayer, a product of Francome Fabrications, is a single, free-standing, compact unit designed for use with both short and long-wheelbase Land Rover Pick Ups Originally designed for control of the Desert Locust 'Hopper' bands by the ultra-low-volume application of insecticides, it has also been found to provide a good method of ultra-low-volume spraying against other pests. Operated by the pressure of the Land Rover's exhaust gases, this 81kg unit is fitted with two 50-litre tanks to allow the operator a choice of two different insecticides. The top-mounted emission jet stack pipe projects fine particles of concentrated chemical into the air, drift technique then taking over. The chemical is blown into the air together with the exhaust gases, thus eliminating the need for an expensive pto pump drive, a pump for liquid chemicals and vulnerable sprayer booms.
(16: Francome Fabrications Ltd, 1994)

16

Aid to development...

Four Defender One Ten Tdi Station Wagon-based mobile cine-video presentation units were supplied to the Ethiopian government by the Dutch-franchised Land Rover dealer Automobielcentrale De Uiver bv in 1992. These vehicles are used as information units, giving advice to the local population about basic issues such as food, hygiene and health, and are fully fitted with all necessary cine and video equipment; dust and shockproof light-alloy storage boxes ensure their survival of bad local road conditions. The unit's equipment includes a generator set, fire extinguishers, a camcorder, Barco video recorder, PA amplifier, one display board, a Barco back projector, toolset, floodlights on tripod, projection screens, external speakers, a 12-volt 750-Watt inverter, an extra battery, a TV monitor, storage boxes for video and cassette tapes, a 16mm Eiki film projector, a radio-cassette player and a Philips mixing amplifier.
(17, 18: Automobielcentrale De Uiver bv, 1993)

17

18

The Land Rover One Ten-based mobile cinema has been developed to present films to the world's more remote communities. This purpose-built unit includes a 16mm film projector with ancillary equipment – which is contained in a lockable cabinet for protection and security – and a portable generator which provides the power and is removed from the vehicle during the film presentation. Since 1970 Agri-Visual have been providing a professional communication service to development organizations such as the International Aid Organization, World Bank, United Nations and European Development Fund.
(19, 21: Agri-Visual Ltd, 1992)

19

20

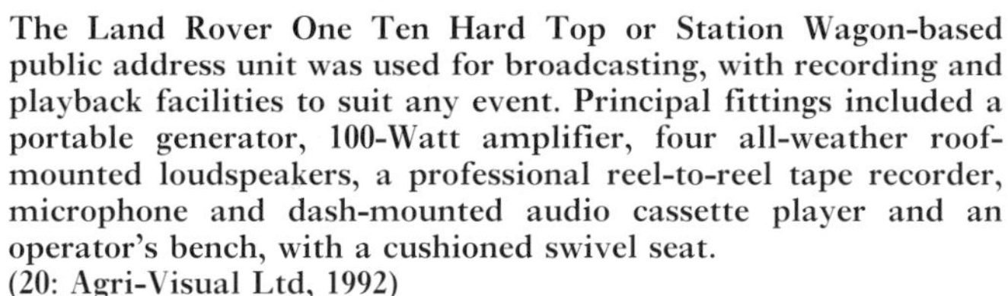

The Land Rover One Ten Hard Top or Station Wagon-based public address unit was used for broadcasting, with recording and playback facilities to suit any event. Principal fittings included a portable generator, 100-Watt amplifier, four all-weather roof-mounted loudspeakers, a professional reel-to-reel tape recorder, microphone and dash-mounted audio cassette player and an operator's bench, with a cushioned swivel seat.
(20: Agri-Visual Ltd, 1992)

21

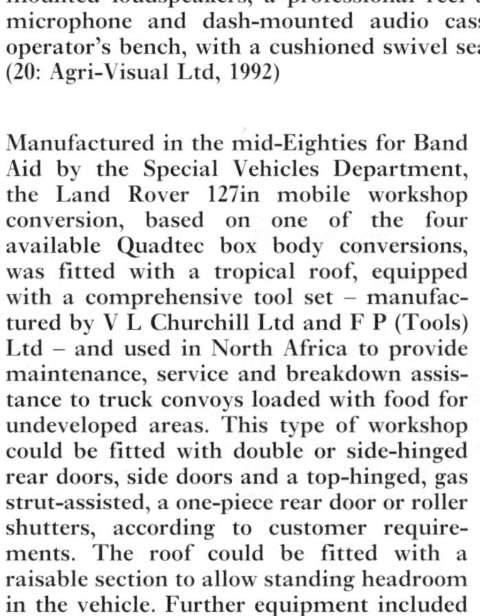

Manufactured in the mid-Eighties for Band Aid by the Special Vehicles Department, the Land Rover 127in mobile workshop conversion, based on one of the four available Quadtec box body conversions, was fitted with a tropical roof, equipped with a comprehensive tool set – manufactured by V L Churchill Ltd and F P (Tools) Ltd – and used in North Africa to provide maintenance, service and breakdown assistance to truck convoys loaded with food for undeveloped areas. This type of workshop could be fitted with double or side-hinged rear doors, side doors and a top-hinged, gas strut-assisted, a one-piece rear door or roller shutters, according to customer requirements. The roof could be fitted with a raisable section to allow standing headroom in the vehicle. Further equipment included a 240-volt electric generator, an air compressor, gas/electric welding equipment, a vice, a workbench and full interior lighting.
(22: Land Rover Ltd, 1985)

22

Airfield operations...

A Series III 109in Hard Top was required for this mobile radar unit conversion for airfield and coastal surveillance. Supplied by Abbey Electronics in 1980, the system could be operated within four to five minutes of arriving on site. This vehicle carried a 12in-display Kelvin Hughes Series A radar set, generator set, switchboard, an earthing stake, a compass, air conditioning, a plotting table, transceiver and rectifier. (23, 24: Kelvin Hughes, 1980)

24

23

25

26

Because of the radar equipment's high position, the Land Rover's original centre of gravity was raised and therefore a new safe-operation transverse angle had to be established, using the Land Rover factory tilt-measuring unit.
(25: Kelvin Hughes, 1980)

The Land Rover's high all-terrain mobility means that it can also be used to find the optimum location for the erection of a new non-mobile radar station. Shown here is the Racal-Decca Solid State '150' radar system option to (23).
(26: Abbey Electronics Ltd, 1980)

27

A Land Rover One Ten Station Wagon aircraft guidance vehicle, seen here in service at Liverpool Airport. It is driven slowly in front of recently-landed aircraft, and is fitted with a multi-message roof-mounted lightbar manufactured by Premier Hazard Systems (UK) Ltd. Through a matrix of fibre optics, it can show the messages: 'Follow Me' in green; 'Slow' in amber; or 'Stop' in red.
(27: Premier Hazard Systems <UK> Ltd, 1993)

28

A Land Rover One Ten Station Wagon-based ground operation vehicle conversion in service at Heathrow Airport. This orange-striped yellow vehicle is fitted with a big glassfibre sign-unit on the roof on which two rotating beacons are fitted to show 'Operations' or 'Medical', depending on the particular duty of the vehicle. A roof-mounted loudspeaker is also fitted and connected to a tape recorder inside, which is used as a bird-scare unit. Hella spotlights are situated above the left and right front doors and are used during night-time runway inspections to trace and remove objects like nuts, bolts and small stones in order to prevent serious damage to jet engines. The vehicle is also fitted with radio equipment for communication with the control-tower. This conversion was carried out by the Heathrow Airport specialist workshop, which has also produced Land Rover One Ten Hard Top-based 'Follow Me' vehicle conversions, of which the Matchbox model car company brought out a 1:37 scale diecast version in 1987.
(28: Heathrow Airport Ltd, 1992)

29

30

The Land Rover Angloco Rapid Response Foam Tender conversion is a lightweight, highly manoeuvrable unit designed to meet the increasing demand for a first-strike appliance, or for where the deployment of a larger vehicle would be impractical. It is particularly suitable for combating flammable-liquid fires occurring at airfields, heliports, oil refineries, storage areas and chemical plants. The appliance is based on the four-wheel-drive Land Rover chassis powered by a 3.5-litre V8 petrol engine. The primary fire suppression system consists of a pressurized cylinder containing 500 litres of pre-mix water-foam solution. The unit is pressurized by standard rechargeable BA cylinders. Discharge from two delivery hoses and foam branches is simple and effective as it is done by valve operation only, and consequently saves vital seconds over conventional foam-pumping appliances; additional support media is provided by two 12kg BCF fire extinguishers, which are readily accessible at the vehicle's rear. The standard appliance comes complete with two side lockers housing the pre-mix unit controls, hoses and foam branches, and has stowage space for additional equipment.
(29, 30: Angloco Ltd, 1992)

31

This Fire Fighting Airfield Crash Rescue vehicle was manufactured to the requirements of the RAF by Foamite Ltd, and was employed for the first time in 1960. A Land Rover 109in Pick Up was used as the basic vehicle, which was employed as a rescue vehicle for crashed aircraft and for extinguishing aircraft wheel-brake fires. Its fire equipment consisted of two dry high-pressure powder extinguishers, one ladder, pneumatic-powered saws, blankets, gloves, a searchlight and site-illuminating lamp.
(31: The Tank Museum, Wareham, 1970)

32

Two Land Rover One Ten-based Series 5 S54 anti-hijack vehicles have been operating since 1986 at Schiphol Airport (Amsterdam) courtesy of a special Dutch police department. Such vehicles were developed specifically for airport reconnaissance, patrol and security operations, and are in service at major airports worldwide. This conversion is fitted with a hardened armoured-plate, extra bulletproof side windows and a specially designed turret with a vision block and ball mount for fitment of a sniper's rifle capable of firing in any direction. The vehicle can be driven to any point on the airfield and positioned close to an aircraft, providing a mobile armoured observation point from which hostile elements could be covered without an overt display of weaponry. Its crew consists of a driver, commander/observer and marksman. In 1969 the Israeli government stipulated the need for security measures against terrorist activities to EL-AL aircraft, passengers and personnel, of which the constant 24-hour use of these airport vehicles was included. Before 1986, six Land Rover 109in-wheelbase Shorts armoured vehicles had rendered the same function. See also (60).
(32: Korps Rijkspolitie Dienst Luchtvaart, 1993)

33

Compact, fast, highly mobile and manoeuvrable light-duty fire support vehicles, based on the Land Rover/Defender One Ten High Capacity Pick Up, were manufactured by Simon Aviation in 1992 for customers in Saudi Arabia and Tanzania. This first strike/rapid intervention vehicle, known as the Fireranger, can be fitted with a pto-driven Godiva waterpump for 2,300 litre/min output and 7 bar operating pressure, a 450-litre capacity glassfibre watertank, 250kg dry powder discharge unit, expellent nitrogen and a 450-litre water/foam premix. Each vehicle is fitted with a heavy-duty suspension and a rear anti-roll bar. Maximum GVW is 3,050kg. (33, 34: Simon Gloster Saro Ltd, 1992)

34

35

Simon Aviation also manufacture two/four-stretcher airfield ambulances and One Ten Station Wagon-based command cars like this example.
(35: Simon Gloster Saro Ltd, 1992)

37

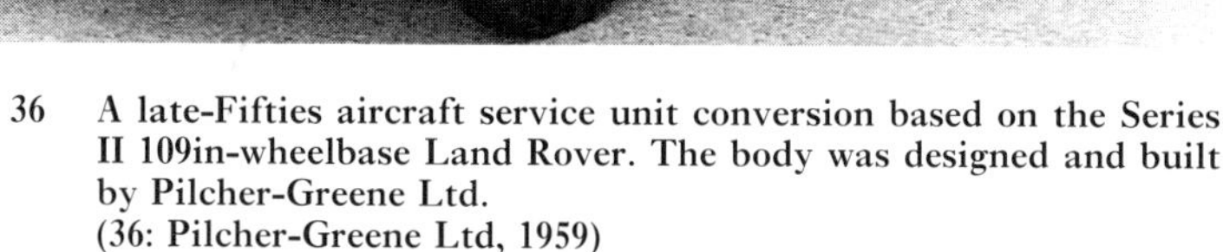
36 A late-Fifties aircraft service unit conversion based on the Series II 109in-wheelbase Land Rover. The body was designed and built by Pilcher-Greene Ltd.
(36: Pilcher-Greene Ltd, 1959)

A Land Rover conveyor conversion driven by centre-pto and manufactured by H W Edghill Equipment Company Ltd during the mid-Sixties. This conveyor consisted of a three-section beam, and was mounted on a long-wheelbase Land Rover Pick Up. It was first used in the coal mining industry for loading lorries, and later several units were used by British Airways for loading passenger baggage into airliners. The top and middle section of this conveyor could have their respective heights adjusted by a front-mounted hydraulic elevating strut, which facilitated loading into different types of aircraft. A specially shaped window was fitted in the roof of the cab for easy observation of the top of the conveyor when approaching the airliner's baggage door.
(37: British Motor Industry Heritage Trust/Rover Group, 1965)

38

V F Specialist Vehicles of Marsden, in Huddersfield, Yorkshire, produce a series of Land Rover-based fire appliances intended for industrial or airfield use. Introduced in 1979, the original model was based on the 109in chassis and featured a front-mounted pump, 120-gallon tank, full-length bodywork and fitted hose reel. The model was updated when the 110in Land Rover appeared, although four pre-production MkII versions were, in fact, tried out on the 109in chassis. Similar in design to the MkI version, the first all-new vehicle came out in 1986, this time using a rear-mounted pto-driven pump. It was equipped with 450-litre water and 60-litre foam tanks. A number of stretched versions of the MkII were also built in order to increase the water capacity, with two of them holding 900 and 1,000 litres respectively. In these instances the models were cut in half and then lengthened to the 6x6 format. All were fitted with a V8 petrol-engine. At the time of writing only the MkII 110in 4x4 version was still available.
(38: Nick Dimbleby, 1993; 39: V F Specialist Vehicles Ltd, 1993)

39

Pyrene water tenders fitted with the same self-priming Coventry Climax pump, for an output of 300 gallons per minute, a control panel, an 80-gallon water tank, a first-aid hose on reel, chromium-plated electric fire bell, ladder and suction hoses. Two body styles were offered – open or closed. During the Sixties, the Pyrene Company Ltd manufactured fire appliances for use with the Land Rover 109in Series IIA. They produced a wide range of water, dry powder and foam tenders, known as Pyrene-Sun Land Rover Appliances, suitable for all fire classifications.
(40, 41: Chubb Group Services Ltd, 1972, 1969)

40

41

42

The Pyrene dry powder unit in its basic form is particularly suited to the smaller airfield. It has two 150lb dry chemical units mounted at the rear of the vehicle, each complete with a fold-flat hose and applicator.
(42: Chubb Group Services Ltd, 1969)

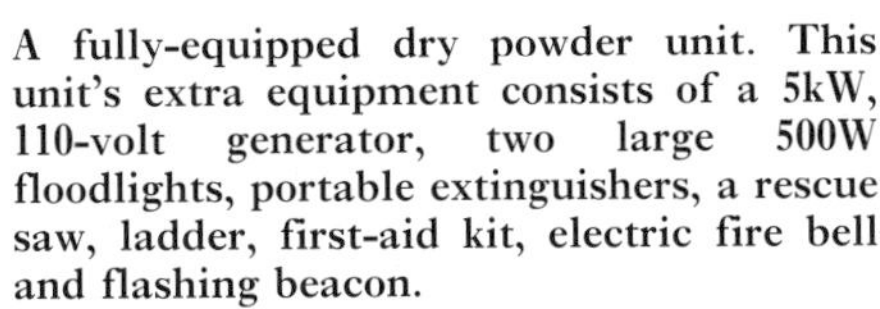

A fully-equipped dry powder unit. This unit's extra equipment consists of a 5kW, 110-volt generator, two large 500W floodlights, portable extinguishers, a rescue saw, ladder, first-aid kit, electric fire bell and flashing beacon.
(43: Chubb Group Services Ltd, 1972)

43

The Pyrene fire appliance on the Land Rover 109in Series IIA was designed to discharge up to 6,126 litres of 'light water' foam. In 1968, hydrocarbon fires in excess of 3,500sq ft were successfully extinguished at the RAF Fire Service Training School, at Manston, by using this method. The 3M company has since developed this unique fire-smothering effect, opening up exciting possibilities in aircraft protection.
(44: Chubb Group Services Ltd, 1969)

44

Armoured vehicles...

45

46

Discreetly armoured Land Rover, Defender, Discovery and Range Rover conversions are manufactured by MacNeillie & Son Ltd and supplied to government agencies, police forces, royal families, state rulers and industrialists – to name but a few! Although visibly similar to standard-specification vehicles, they offer defence against high-velocity rifles firing military ammunition. Such protection may consist of a fully-armoured steel panel body including floor-pan and roof, glass/polycarbonate-armoured windows, separate protected fuel tanks and batteries, a heavy-duty suspension system, run-flat tyres and an automatic fire extinguisher system – depending on customer specification.
(45, 46: MacNeillie & Son Ltd, 1993)

47

48

Some customers required an armoured cash-in-transit vehicle with the facility for long-range operation. Glover Webb Ltd were able to fit an additional tank to the rear of the body, which was fed into the rear fuel tank when required. Full air conditioning was supplied to the body and cab from the roof-mounted unit operated from the vehicle's engine.
(47, 48: Glover Webb Ltd, 1980)

This armoured Land Rover 109in Hard Top was supplied in 1980 for operation in Yemen. Each of these vehicles was fitted with a Hubbard Engineering air conditioning system.
(49: Glover Webb Ltd, 1980)

49

The Land Rover 109in-wheelbase Hard Top makes an affordable small mobile bank with one pay-out hatch and full body and cab air conditioning. This particular unit was destined for the African market in 1980, but Glover Webb Ltd still offer such a conversion for use worldwide. This type of vehicle has full protection against 9mm automatic gunfire. All windows are made of laminated 30mm bulletproof glass which provides the same standard of protection as the steel body armour. Run-flat tyres and extra fuel-tank protection are also fitted.
(50: Glover Webb Ltd, 1980)

50

With the introduction of the V8 petrol engine, customers were able to obtain a fast armoured cash-in-transit vehicle. This example has air conditioning and an interlocking transfer hatch fitted into the rear door. This type of conversion is still offered on the Land Rover One Ten with the choice of petrol V8 or turbo diesel engine.
(51: Glover Webb Ltd, 1991)

51

This fully armoured Land Rover 109in Station Wagon was purpose-built for the King of Nepal after his coronation in 1974. It was finished in black, had red leather trim and was installed with full air conditioning. The rear seats, retrimmed to match, were moved back to enable the King to stand up through the armoured roof-mounted hatch.
(52: Glover Webb Ltd, 1974)

52

53

One of a small fleet of armoured 109in-wheelbase Land Rover Station Wagons built for the Indonesian government in 1979. This fully air conditioned vehicle could carry the driver and one guard in the front and two guards in the rear; the centre section was partitioned-off to accommodate two VIPs. A spotlight was fitted within the cab roof and a Warn self-recovery winch was attached to the front bumper.
(53: Glover Webb Ltd, 1979)

54

An armoured Land Rover One Ten-based mobile bank conversion for a Tanzanian customer, manufactured by Glover Webb Ltd. Protection is rendered to the same specifications as provided in the company's cash-guard conversions.
(54: Glover Webb Ltd, 1991)

55

Dutch manufacturer ITM (International Transport & Management) in Doesburg specializes in supplying armoured Land Rover Defenders to customer requirements. A special type of armoured steel sheet is used throughout the cabin in order to provide protection up to level M4 (DIN). The vehicle is designed to withstand 7.62mm NATO ball ammunition from a distance of 10m at an angle of 90deg (speed approximately 850m/sec). Window panels are made from 39mm thick multi-layer glass with polycarbonate, providing protection up to level C4 (DIN) – equal to the armoured steel protection. Special backings are also used, made from Aramid Fibers (Twaron), which gives additional protection against mines and steel shivers. All the vehicles are equipped with run-flat tyres, and because of the additional weight incurred (approximately 700kg) heavy-duty suspension with double coil springs is fitted. It takes ITM only four weeks to convert a standard Rest of World-specification Land Rover Defender 110 Hard Top into a fully armoured vehicle, some of which were supplied to the news teams of broadcasting corporations and others to relief organizations operating in the Balkan war.
(55: International Transport & Management, 1994)

56

A white-painted Shorts Shorland armoured vehicle supplied by Transport Resource Management Ltd in 1992 to BBC news teams during the Balkan war.
(56: Transport Resource Management Ltd, 1993)

Shorts-manufactured Shorland range of armoured vehicles are primarily designed for internal security roles and are in service in over 40 countries. The Land Rover 109in-based armoured patrol car has proved to be a cost-effective answer to terrorist activity in urban and rural areas and to be very useful in situations of civil unrest. It carries a crew of three and may also be used for border control, convoy escort, reconnaissance and internal security duties. Smoke projectors and a Browning machine gun are popular options for this conversion.
(57, 58, 59: Fotografie Lex Klimbie, 1989)

57

58

59

The S54 anti-hijack vehicle, also shown on airport duty in (32). This, one of the Shorts Series 5 vehicles, is based on a specially strengthened Land Rover One Ten chassis-frame and powered by the Rover V8 3.5-litre petrol engine, coupled to permanent four-wheel drive. Their armour has proved effective against both 7.62mm rifle and machine gunfire, even at very close range. Their floor is constructed from tough GRP, providing protection against blast, nail and pipe bombs.
(60: Short Brothers plc, 1990)

60

The S55 armoured personnel vehicle, developed for the safe and rapid transportation of security forces in high-risk areas, is designed to provide seating for an eight-man crew. It has eight gun ports, and laminated glass inserts – three on each side of the crew compartment and one in each rear door – for the use of sub-machine guns. A roof-mounted machine gun hatch and electrically-operated smoke grenade dischargers are available as optional extras. Burning fuel from petrol bombs thrown against the windscreen is diverted directly to the ground – without entering the engine compartment – via a special gutter positioned under the windscreen.
(61: Short Brothers plc, 1990)

61

62

The S52 armoured patrol car. This conversion provides a cost-effective aid to law enforcement, both cross-country or on the streets. It has a crew of three: driver, commander and a gunner in the fully-rotating manually-operated turret, which may be fitted with many types of armament including the NATO 7.62mm general-purpose machine gun. Electrically-operated smoke grenade dischargers are an optional extra. The windscreens are protected by drop-down armoured visors with laminated glass inserts; entry/exit is provided by the front doors, two hatches in the turret and a hatch at the rear of the vehicle. This conversion's unladen weight is 2.81 tonnes, and laden 3.13 tonnes.
(62: Short Brothers plc, 1990)

63

The S53 air defence vehicle is based on the S52 armoured patrol car and has been developed to provide effective, low-level air defence over strategic areas. It is fitted with a variant of the Blowpipe/Javelin Lightweight Multiple Launcher (LML).
(63: Nick Dimbleby, 1993)

64

65

Light-armoured Courtaulds Aerospace CAV100 vehicles, based on the 3.6-tonne GVW Land Rover Defender chassis, purchased in 1993 by the United Nations High Commissioner for Refugees (UNHCR), to protect relief workers in Bosnia. A range of configurations was available, suitable for military logistics and reconnaissance, peace-keeping, internal security, ambulance and communication roles. Courtaulds Aerospace Advanced Material Division manufactures a range of armoured vehicles and vehicle protection systems, relying exclusively on lightweight fibre-reinforced composite armour.
(64, 65: Courtaulds Aerospace Ltd, 1993)

66

67

The CAV100 was also ordered by the BBC for their news teams operating in Bosnia. Transport Resource Management Ltd supplied them the vehicle in December 1992, but the temporary export of this conversion to the Balkans needed special permission and licences from the Department of Trade and Industry because of the sanctions on exports to Serbia and surrounding countries at the time. The outstanding level of protection offered by the CAV100 against small arms and mines had never before been achieved by vehicles at this weight and payload.
(66, 67: Transport Resource Management Ltd, 1993)

68

A Land Rover One Ten Hard Top, converted by S M C Engineering (Bristol) Ltd to a fully armoured vehicle and supplied to television news teams such as ABC, BBC, CBS and the Reuters agency, for operation in areas where armoured protection is most needed.
(68: S M C Engineering <Bristol> Ltd, 1993)

Body conversions (general)...

A Land Rover Discovery Hi-Roof conversion carried out by the Dutch company Terberg Specials bv to change a Discovery car into a van. In Holland the rate of road tax for a van is approximately 60 per cent of that for a car. However, recently it became compulsory to blind these vans' rear side windows – even when fitted with the Hi-Roof.
(69: Terberg Specials bv, 1993)

69

70

An elegant and characteristic raised-roof body, built in series production by E V Engineering Ltd and used for many applications where extra vertical space is needed. The roof of this body is rigidly constructed from a welded heavy-gauge aluminium panel.
(70: E V Engineering Ltd, 1993)

71

The Scott Broadcast Land Rover with a wheelbase of 120in, based on the Land Rover One Ten Station Wagon and used by the BBC. The chassis and body of this vehicle has been extended by 10in, allowing two new narrow doors to be fitted between the original doors. However, the raised roof panel has remained at its original length.
(71: Nick Dimbleby, 1993)

72

73

POD Ltd manufacture a demountable GRP body, designed around the Land Rover 110in Hi Capacity Pick Up. They progressed from this to produce a variation which allows them to mount a POD onto the 110in general service Land Rover as used by the British Ministry of Defence. This demountable body unit's versatility is reflected in its recent usage as a civilian ambulance, military ambulance, mobile workshop, camper, and command and communications centre, and has been sold worldwide. The units shown are performing military ambulance roles for the Kuwait government. The body may be secured to the vehicle in four places in a matter of minutes. If fitted as an ambulance it is capable of carrying either two prone patients and an attendant, or up to eight seated patients plus attendant. The interior configuration or function may also be altered by the inclusion of modules, which can be fitted in seconds. The unit, which weighs approximately 300kg, can be fitted with an electrically-pumped water system complete with 20-litre storage tank, a removable basin and waste receptacle, two 15-Watt air blowers and an emergency window exit with quick-release fittings. The side panels are insulated by foil and glass wool and the roof panels by expanded foam. A roof-mounted air conditioning system can also be fitted if required. Satisfied customers have included such organizations as UNICEF, ECHO, UNIDO, St John Ambulance, International Red Cross, Red Crescent, Trimed and Wellfind.
(72, 73: Shanning POD Ltd, 1992)

74

This heavy-duty Land Rover Pick Up 6x6 conversion by Reynolds Boughton Ltd, based on the One Ten chassis, was designed and developed to meet an increasing requirement for a high-powered cross-country vehicle with a large passenger and cargo-carrying capacity. A feature of the 6x6 is that the third driven axle improves the vehicle's cross-country ability.
(74: Reynolds Boughton Ltd, 1989)

75

A Land Rover 6x6 body conversion being used as a bomb disposal unit and carrying a tracked remote-controlled bomb defuser in the back, known as the Wheelbarrow, a product of Morfax Ltd. It was built for a foreign government by Glenfrome Engineering Ltd.
(75, 76, 77: Bristol & West Photography, 1990)

76

77

78

79

Rear section and roof of a Range Rover replaced by those of a Land Rover Discovery, providing much more space inside the vehicle. Originally the Range Rover and the Discovery had the same 100in-wheelbase chassisframe, so fitting the Discovery's rear body section to the Range Rover was a straightforward task. However, streamlining the former's rear body section to that of the latter's rear side doors proved more difficult. A specially made loadspace trim is fitted to match that of the Range Rover's interior. At the time of writing nine of these conversions had been carried out.
(78, 79: Nick Dimbleby, 1993)

80

One of the very successful Quadtec range of box body conversions for the Land Rover, developed and manufactured by Land Rover's Special Vehicles and on sale since the late Eighties. There are four body configurations, mainly based on the Defender 130in chassis: the raised box body with crew or single cab, and the normal box body with crew or single cab. These aluminium box-bodied vehicles are used as mobile workshops, cinemas, canteens, TV production units, ambulances, aid organization vehicles, communication centres for the military, police and fire brigades and as a Camel Trophy support vehicle.
(80: Land Rover Ltd, 1993)

81

Special box body manufactured and fitted to the Land Rover 130 Chassis Cab by Marshall of Cambridge (Engineering) Ltd, for the armed forces of Oman.
(81: Marshall of Cambridge <Engineering> Ltd, 1990)

82

A Handicraft mobile workshop conversion based on the Series IIA 109in chassis and manufactured by Pilcher-Greene Ltd in the late Sixties.
(82: Pilcher-Greene Ltd, 1969)

83

A special 12in roof extension for the Defender One Ten Station Wagon and Hard Top manufactured by Macclesfield Motor Bodies (UK) Ltd and weighing approximately 40kg. It is fully trimmed and can be fitted with fixed and/or opening side windows. If the original Defender roof is retained, the extension can be used as a stowage area, for which side and/or rear access doors are fitted. Surveillance-vehicle conversions use this stowage area to carry electronic equipment.
(83: Macclesfield Motor Bodies <UK> Ltd, 1990)

84

85

A demountable multi-purpose GRP Pick Up canopy, manufactured by the Dutch company Fabuglass Europe bv and available since January 1993 exclusively from the Dutch Land Rover distributor/dealer organization. The canopy, which weighs 31kg with standard fittings, is engineered to fit the Defender 130 Crew Cab Pick Up and its exterior height, width and length are 0.86 x 1.70 x 1.74m respectively. Only available in white for the exterior and grey for the interior, the canopy is attached to the Defender Pick Up body by nuts and bolts and includes a framed, gas-strut-assisted window upper tailgate, which fits to the Defender's existing lower tailgate. Interior lighting is standard, and the canopy roof is fitted with special mountings to provide a solid attachment for a roof-rack.
(84, 85: Cars Gravemeijer, Rover Nederland bv, 1993, 1992)

86

These large flat-bed dropside body conversions on a Defender 6x6 Chassis Cab, which are engineered by Special Vehicles, are produced in volume by P D Stevens & Sons Ltd, who have been involved in Land Rover conversion work for more than 30 years. Some of these high-payload conversions, with their excellent cross-country capabilities, are used by water laboratories for field work. The dropsides are made from hollow-section extruded aluminium.
(86: P D Stevens & Sons Ltd, 1993; 87: Land Rover Ltd, 1993)

87

A Land Rover Forward Control conversion based on the Stage I 109in-wheelbase V8 – note the headlight panels, grille and bonnet lock. The chassis was developed by SMC Engineering (Bristol) Ltd – formerly known as Sandringham Motor Company, who have specialized in Land Rover conversions for more than 30 years – and built by Hotspur Engineering Ltd. The flat-bed (not shown) was made by Longwell Green Coachworks Ltd, who were later taken over by W H Bence Ltd. The standard all-aluminium dropside flat-bed body measured 2.54 x 1.78m and gave a payload of 1.3 tonnes, while a similar body fitted to a permanent six-wheel-drive version of this vehicle – with a wheelbase of 139in – had a 4m long bed and a payload of 2 tonnes. Examples of this vehicle – known as the FC82 – with various bodies were sold to British water and electricity authorities and to the British Army; one example was sold to the Dutch Army for trial purposes. It was suitable for use as a light artillery gun tractor, ambulance, field command post, personnel carrier and radio communication unit. However, only 500 of these Forward Control Land Rovers were built, from 1982–85.
(88: SMC Engineering <Bristol> Ltd, 1982)

88

More examples of bodywork conversions by P D Stevens & Sons Ltd. Here is an alloy rear body with electro-hydraulic tipping gear mounted on the 109in-wheelbase Chassis Cab.
(89: P D Stevens & Sons Ltd, 1979)

89

This is a wooden stakeside body fitted on a Series IIA Land Rover 109in Chassis Cab, which had a payload of 816kg. The load floor of the dropside version, which could be in ribbed alloy or smooth wood planks, was fitted to two channel members which in turn were fitted to the existing Land Rover chassisframe body mountings. Resin-bonded, hard-face plywood floors were also available for this body, which had a flat cargo floor space of 2.03 x 1.68m. The dropsides and tailboard were made of 15in deep, heavy-duty alloy pressings fitted with mild steel hinges and fittings. Rope hooks, protected rear lamps and detachable corner pillars were standard fitments. This body could also be supplied in steel, and a ladder gantry was another option. The heavy-duty rear suspension included hollow rubber springs and uprated shock absorbers were fitted all round.
(90: P D Stevens & Sons Ltd, 1965)

90

91

92

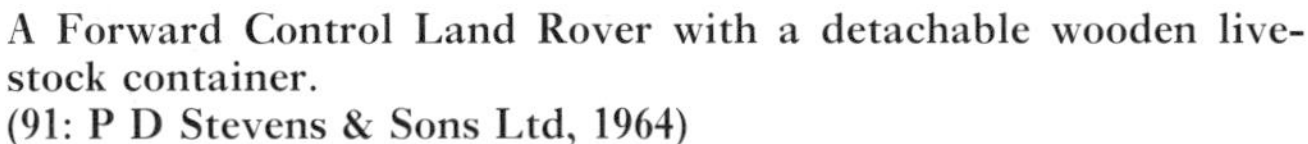

A Forward Control Land Rover with a detachable wooden livestock container.
(91: P D Stevens & Sons Ltd, 1964)

An electric crane mounted on a Stevens alloy dropside body conversion for the Land Rover 110in Chassis Cab.
(92: P D Stevens & Sons Ltd, 1984)

93

Ifor Williams Trailers Ltd, in addition to manufacturing a comprehensive range of high-quality trailer products, also produce a number of pick up canopies of various sizes for the Ninety, One Ten and High Capacity Pick Up. This type of canopy is constructed from sturdy aluminium/galvanized steel and can be fitted with either an open-weld mesh upper tailgate or a closed one panelled in aluminium.
(93, 94, 95: Ifor Williams Trailers Ltd, 1993)

94

95

96

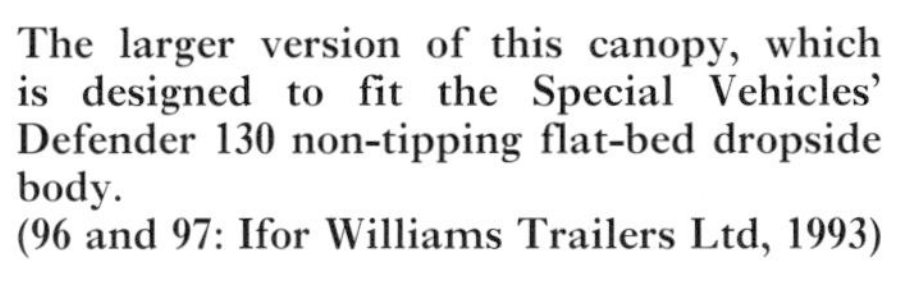
The larger version of this canopy, which is designed to fit the Special Vehicles' Defender 130 non-tipping flat-bed dropside body.
(96 and 97: Ifor Williams Trailers Ltd, 1993)

97

99

Known as the Truckman Top, these demountable GRP pick up canopies are designed specifically to fit the One Ten High Capacity Pick Up and the 130 Crew Cab Pick Up. The multi-purpose canopies have an insulated, double-skinned sandwich roof and are available with or without large opening side windows. The rigid upper tailgate is fitted with high-pressure gas struts for easy opening. The interior height, length and width of these canopies is 1.41 x 1.97 x 1.63m (for the One Ten High Capacity Pick Up) and 1.41 x 1.66 x 1.63m (for the 130 Crew Cab Pick Up); their respective weights are 59kg and 49kg. (98, 99, 100 and 101: Truckman Ltd, 1993)

98

100

101

The first Land Rover Station Wagon was completed in July 1948, being built on the Land Rover 80in pre-production rolling chassis by the coachbuilding company Tickford, of Milton Keynes, Bedfordshire. The body was constructed on an ash frame, and fitted out as a seven-seater. About 640 of them were built from 1948–51, and to meet the demand during 1949–50 the London-based coachbuilders Mulliner – later to become part of Rolls-Royce – were also subcontracted to this work. The Station Wagon was priced very expensively on the UK market and as a consequence the vast majority of them were exported. This vehicle featured permanent four-wheel-drive transmission, being provided with a freewheel in the front driveline and two constant-velocity universal joints in the front axle. To be strictly accurate, the Land Rover Station Wagon was not a conversion, but a standard factory body option.
(102: Cliff Petts, Newport Pagnell, 1950)

102

This aerodynamically-shaped GRP High Top conversion, based on a Series III Land Rover 109in Hard Top, was manufactured by Marshalls of Cambridge for a Danish customer.
(103: Marshall of Cambridge <Engineering> Ltd, 1978)

103

A rare Series III Land Rover 109in V8-based conversion, also by Marshalls, showing the unusual combination of a station wagon and a soft top, which was made on special request for a Middle East customer.

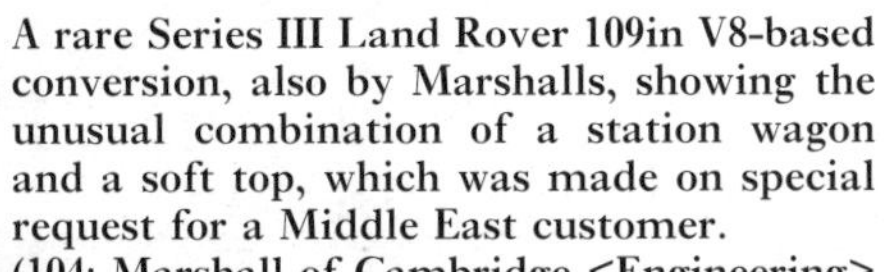

(104: Marshall of Cambridge <Engineering> Ltd, 1982)

104

105

An early State Review Vehicle based on the Series I short-wheelbase Land Rover. Owned and used by the British Royal Family during the Fifties and Sixties, the specially designed body and conversion work was done by Land Rover Ltd.
(105: National Motor Museum, Beaulieu, 1956)

106

A Land Rover One Ten with heavy-duty chassis, which was converted for the use of the Pope by Glover Webb Ltd. The whole body was air conditioned and lined in blue, and a small seat was fitted. Steps were provided for ease of access and there were grab rails at the rear and sides for the guards. It was built in 1989 and shipped out to Mauritius for the Pope's visit.
(106: Glover Webb Ltd, 1989)

108

107 A hearse conversion based on one of the last Series IIA 109in Land Rovers, with the old-style grille and new-style headlight panels; it was one of 25 manufactured by Pilcher-Greene Ltd for a Libyan customer. Fully air conditioned, it had a wooden interior with a raised centre section into which four rollers were fitted for easy loading of the coffin. To the left and right of this elevated centre section seats were provided for the bearers.
(107, 108: Pilcher-Greene Ltd, 1971)

109

A viewing vehicle produced by Pilcher-Greene Ltd for the visit of HRH Queen Elizabeth the Queen Mother to game parks in Sierra Leone. Finished in green, this vehicle had large side windows and facilities for a sliding hatch on the roof with camera-tripod mounts as well as a raised floor to enable their operation from inside the vehicle. The vehicle's interior was finished almost entirely in leather, with special fabric inserts in parts of the seats.
(109: Pilcher-Greene Ltd, 1959)

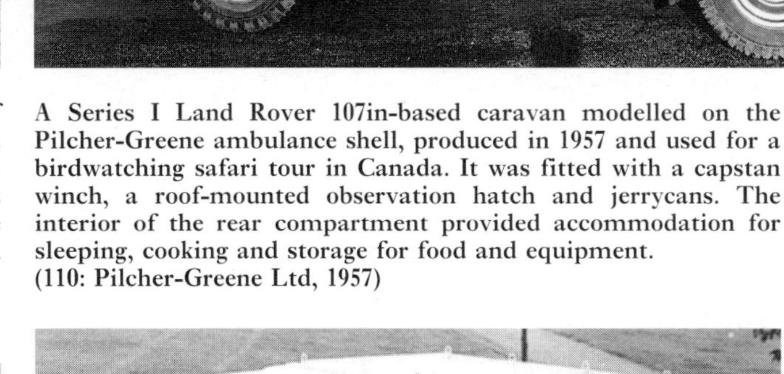

110

A Series I Land Rover 107in-based caravan modelled on the Pilcher-Greene ambulance shell, produced in 1957 and used for a birdwatching safari tour in Canada. It was fitted with a capstan winch, a roof-mounted observation hatch and jerrycans. The interior of the rear compartment provided accommodation for sleeping, cooking and storage for food and equipment.
(110: Pilcher-Greene Ltd, 1957)

111

A review vehicle based on the Land Rover One Ten Hard Top, designed and built on request for an African government by Special Vehicles.
(111: Land Rover Ltd, 1992)

112

A canopy manufactured and fitted to a Series IIA Land Rover 109in Pick Up by Sambell Engineering Ltd. This canopy was made of Zintec metal and was also available for the Land Rover 88in Pick Up. An extra wide top-hinged tailgate with window and built-in roof-rack came as standard fitments; side windows were optionally available. Sambell Engineering Ltd manufactured Land Rover Pick Up canopies until 1984.
(112: Sambell Engineering Ltd, 1970)

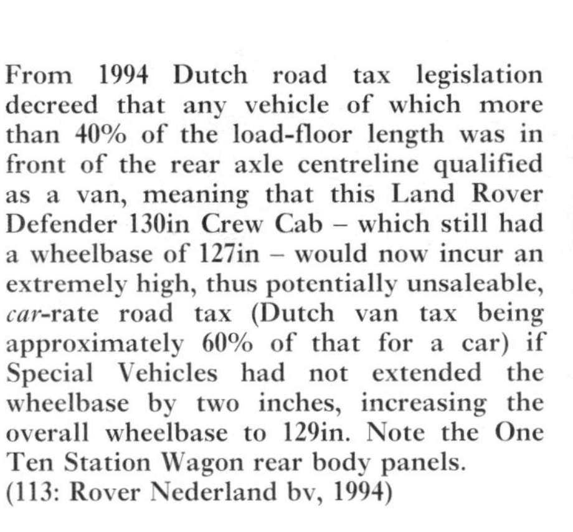

113

From 1994 Dutch road tax legislation decreed that any vehicle of which more than 40% of the load-floor length was in front of the rear axle centreline qualified as a van, meaning that this Land Rover Defender 130in Crew Cab – which still had a wheelbase of 127in – would now incur an extremely high, thus potentially unsaleable, *car*-rate road tax (Dutch van tax being approximately 60% of that for a car) if Special Vehicles had not extended the wheelbase by two inches, increasing the overall wheelbase to 129in. Note the One Ten Station Wagon rear body panels.
(113: Rover Nederland bv, 1994)

Civil engineering and road building...

114

115

116

Designed especially for the Defender One Ten Chassis Cab-engineered hydraulic digger is the Beaver D270 back-hoe, manufactured by Hayters plc since 1987. It is a very fast, economical, all-terrain digger for light work to a depth of 2.10m. In addition to other special-purpose attachments, trenching buckets from 22–60cm width and a hydraulic post-hole borer attachment for 23cm and 30cm diameter boring work are also available. The hydraulically-operated oil pump is driven by the power take-off. These go-anywhere back-hoe diggers are used by the military, geological and construction operators and municipal workers. (114, 115, 116: Hayters plc, 1991)

The Hands-England Division of James Howden & Co Ltd has manufactured several different types of Land Rover-mounted drill rigs. This type, known as the Drillmaster Minor, was mounted on a late Series I Land Rover 109in Chassis Cab during the early Sixties. It had a drilling depth of approximately 30m and was driven by a Norman engine supplied with a Dawson and Downie pump.
(117: Hands-England Division of James Howden & Co Ltd, 1961)

Another type, the Drillmaster Major, is mounted here on a Series II Land Rover Pick Up. This rig was driven by a Volkswagen engine, to drill up to 60m. It was in production during the early Sixties.
(118: Hands-England Division of James Howden & Co Ltd, 1961)

117

18

This is the impressive Drillmaster 18, mounted on a Series III Land Rover 109in Pick Up. Fully hydraulically powered, it is fitted with a hydraulic cylinder for folding the mast into the travelling or working position, another hydraulic cylinder with steel wire assembly positioned in the centreline of the mast to hoist and feed the rotary head and auger string, and a rotary head driven by two hydraulic motors. Mechanical screw-down stabilizing jacks are fitted to the Land Rover's rear crossmember. This rig has a hoist capacity of 1,800kg and its rotary head can develop a torque of 2,300Nm. It is suitable for drilling to a depth of approximately 117m, using a 9cm diameter drill pipe.
(119: Hands-England Division of James Howden & Co Ltd, 1980)

119

120

The latest type, known as the HE 20 LR, shown in the travelling position and mounted on the Defender One Ten Chassis Cab. This is fitted with an electric welded pressed-steel mast which is raised from the travelling position by a hydraulic cylinder. A long-stroke hydraulic cylinder is built into the mast and coupled to the cradle by a wire rope system passing round large rope sheaves, providing hoist and feed capacities of 2,000kg and 1,130kg respectively. The rotary head – mounted on an opening cradle so that it can be swung to one side for handling casing or sampling when a winch is installed – can develop a torque output of 2,000Nm and incorporates a hydraulic motor to drive the main spindle through reduction gears. The hydraulic system is powered by gear-type hydraulic pumps mounted to the centre power take-off. This rig can drill to a depth of 130m with 9cm diameter drill rods.
(120: Hands-England Division of James Howden & Co Ltd, 1991)

121

22

123

A multi-purpose conversion for the Defender 130in, which can be used for five different applications, has been manufactured by Will Engineer Ltd and is known as the Multi-Loader. It is capable of being operated single-handedly for palletized loads, waste disposal skips (collect/drop and tipping action), bowsers and even a hydraulic motor-driven mixer. The vehicle retains its conventional use as a Pick Up, with the added advantage of dropsides and tailboard. The pivoted lifting device is attached to a steel deck support frame. Hydraulic stabilizers are mounted at the rear of the vehicle, the hydraulic power deriving from the pto-driven pump. The hydraulic mixer appliance is easily charged from the standard ready mix or batching plant. The tipping action is used for emptying the mixer on site.
(121, 122, 123, 124, 125: Will Engineer Ltd, 1990)

124

125

126

127

A demountable welded steel tray with hinged tailgate, used in the Sixties to render two functions for a standard short-wheelbase Land Rover: enlarging its load capacity and converting it to a hand-operated tipper vehicle. Before the tipper tray was fitted, the standard Land Rover's lower tailgate had to be removed, its latches and hinges then being used to attach the tipper tray to the body.
(126, 127: F W McConnel Ltd, 1965)

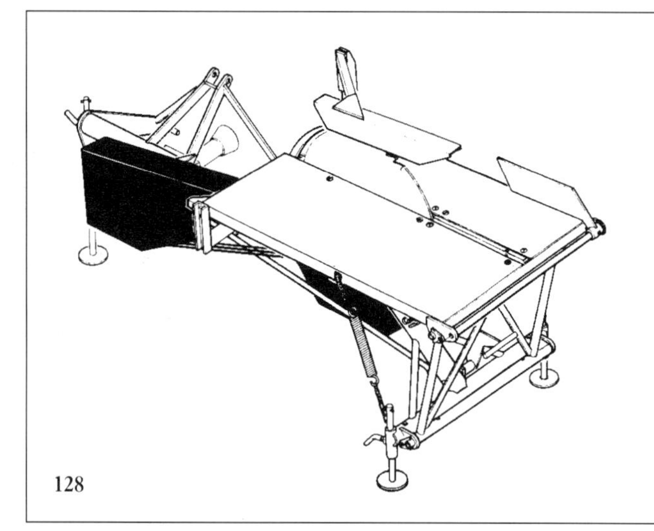

128

A pto-driven all-work sawbench, manufactured by F W McConnel Ltd and used for the first time in the Sixties, when it was attached to a Series IIA 88in Land Rover for agricultural and forestry work. During transit it was lifted by a tractor-type three-point linkage. Very few sawbenches were fitted to Land Rovers due to the rarity of a pto driveshaft and three-point linkage at the vehicle's rear, also because the sawbench was offset to the side to allow, for example, for the cutting of long poles, which meant that it protruded during transportation from site to site. This application also required a steady engine rpm, a large engine flywheel being more suitable for maintaining constant speed with varying loads when cutting. In the early Fifties, E Allman & Co Ltd (Farm Equipment) manufactured a portable sawbench which could be fitted to the rear of the Series I Land Rover. It was fitted with a 24in diameter saw – which could cut logs of 9in diameter and rip timber 2.25m long – and was driven by the vehicle's rear belt pulley. During transit it was folded forwards on a support frame mounted to the Land Rover's rear crossmember.
(128: F W McConnel Ltd, 1965)

129

This is a prototype of the Forest Rover, which was built in 1959 and offered a payload of 1 ton. It was subsequently fitted out as a fire tender.
(129: Forestry Commission, 1959)

130

This strange-looking, 109in Series II Land Rover conversion for forestry work was engineered by the Forestry Commission Experimental Establishment and several production models were manufactured by Roadless Traction Ltd in 1962. Called the Forest Rover, it had very wide axles giving a 13:1 reduction and manufactured by Kirkstall Ltd of Leeds, and big tractor wheels fitted with 10 x 28in Firestone tyres. Its enormous underbelly clearance resulted in extremely good all-terrain capability, which was necessary in order to drive over rough ground surfaces strewn with fallen tree trunks. Because of its big unbalanced wheels and high centre of gravity, on-road performance was very poor, especially at high speed.
(130: Nick Dimbleby, 1993)

131

Mobile workshop conversions for on-site and off-road situations were made for long-wheelbase Land Rover Defenders by Tooley Electro Mechanical Co Ltd. A side work-bench and opening canopy, with a reinforced workbench at the rear in place of the tailgate, is fitted to all their workshops. The impressive range of tools and equipment in this mobile workshop includes: a pto-driven 15kW, 230-volt generator, stowage lockers, an engineer's vice, a heavy-duty drill, drill stand, double-ended bench grinder, three-core cable reel, portable floodlight on tripod, fire extinguisher, metal toolbox containing assorted handtools, complete arc-welding kit, air compressor, tyre inflator and pressure gauge, air-operated grease gun and a gas welding and cutting set. Every workshop supplied is fitted according to the customer's specific requirements.
(131, 132, 133: Tooley Electro Mechanical Co Ltd, 1980, 1980, 1990)

132

133

134

This Defender One Ten pole carrier conversion is fitted with pole erector equipment in the rear and is running on Goodyear Terra-Tyres to provide good traction on soft ground.
(134: Nick Dimbleby, 1992)

135

A sketch of the well-equipped mobile welding plant from the early Sixties manufactured by Lincoln Electric Co Ltd for Series I, II and IIA Land Rovers, and used here to repair farm machinery. Its framed generator was cleverly installed using the hinges and latches from the removed lower tailgate. The V-belts between the rear power take-off and generator were tightened automatically while this installation took place. The required extras for this conversion included a centre pto, an engine-speed governor, an oil pressure and water temperature gauge plus an oil cooler and oil temperature gauge. The first Land Rover-based mobile welders were supplied in 1948.
(135: Land Rover Ltd, 1966)

136

A Defender 130 box body conversion equipped for electrical engineering work on the National Grid. It was produced by Land Rover's Special Vehicles division.
(136: Nick Dimbleby, 1992)

137

138

Hydraulic platforms, usually powered by an engine-driven pump and manufactured by Simon Engineering Dudley Ltd, have been installed on 109in Land Rover Pick Ups since the Sixties for overhead working. All movements of this 8m high unit are controlled by the operation of hand-levers fitted to the operator's cage. Vehicles intended for road use only are provided with rear spring lock-out jacks, but outrigger jacks are provided for cross-country usage. The cage itself is electrically insulated against 1,000 volts and the control valves are protected against accidental contact. Hydraulic-lock valves are fitted directly to all hydraulic cylinders to prevent sudden descent after hydraulic system failures. The cage can be lowered after engine stalling by draining the hydraulic cylinders by means of a bleed valve in the cage.
(137, 138: Simon Dudley Ltd, 1978)

139

140

Powered Access Ltd manufacture Land Rover Defender hydraulic work platform conversions for the Series III High Capacity Pick Up, Defender 130 and Defender 6x6 Chassis Cab. The latest conversion has a GVW of 4.5 tons and a maximum working height of 9.5m. This unit has four hydraulic stabilizers, a maximum outreach of 4.6m and a horizontal rotation of 360deg. Its glassfibre cage is insulated to 2kV. Such Defender conversions are used for overhead inspection and maintenance work to buildings, viaducts, lights and so forth.
(139, 140: Land Rover Ltd, 1992; 141, 142: Powered Access Ltd, 1992; 143, 144: Nick Dimbleby, 1992)

141

142

143

144

145

146

In 1975 Raydel Engineering Ltd took over from Air Drive Ltd the work of fitting air compressors in Land Rovers. Rotary-vane compressors were installed instead of screw compressors, and to date the company remains an approved specialist manufacturer in this field. Shown here is a six-wheel-drive Defender Chassis Cab with an underfloor screw compressor (right) and 110-volt alternator (left), their respective capacities being 100cu ft per minute at 100psi (2.8cu m per minute at 7 bar) and 4kW. Their fitment required the centre power take-off to be modified to take a step-up gearbox capable of producing 4,600rpm with an engine speed of 2,000rpm on full load – sufficient for the required output from the compressor. The speed from the compressor to the generator is reduced with a belt-drive to give the alternator a speed of 3,000rpm. All functions are automatic, once the pto has been engaged, the engine speed being controlled by an electronic management system. A recent use of these expensive conversions was by a water company to attend to bore holes and overland water pipe breakages.
(145, 146: Raydel Engineering Ltd, 1993)

147

Broom & Wade Ltd produced WR100M Land Rover mobile air compressors, which were sold in large numbers worldwide during the Sixties. This is a prototype unit, which was produced in 1961. It has a capacity of 100cu ft per min (2.84cu m) at 7 bar pressure, and went into series production in 1962.
(147: Compair Holman Ltd, 1961)

148

The Broom & Wade production-model mobile air compressor, using an underfloor rotary-vane compressor driven by the centre power take-off and driveshaft. The compressor is efficiently cooled by a recirculatory oil system, the cooler for which is situated in front of the engine's water-cooling radiator. Broom & Wade Ltd produced this conversion until 1969, when it was sold to Air Drive Ltd, who continued the production until 1975, when it was passed to Raydel Engineering Ltd.
(148, 149: Compair Holman Ltd, 1962, 1967)

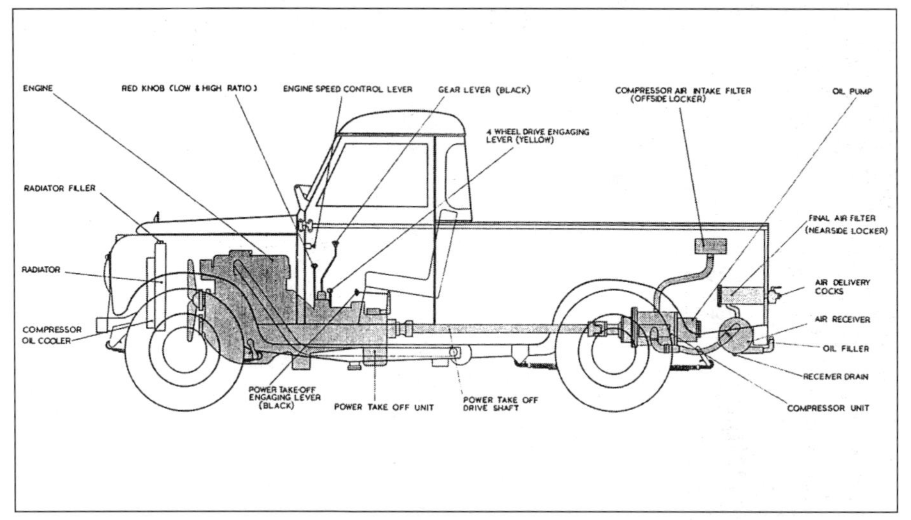

149

In 1949 Alfred Bullows & Sons Ltd converted the Series I Land Rover 80in to a mobile air compressor unit by fitting – via two V-belts – a rear pto-driven rotary Hydrovane air compressor in the back. This was mounted in a steel frame, with hinges and latches to match those of the standard Land Rover tailboard so that when the tailboard was removed, the 140kg compressor could be attached in 30 seconds by two men. At 2,400rpm it had a capacity of 60cu ft per minute (1.7cu m) at 7 bar pressure. The speed control was fitted in the engine linkage, which was operated from the compressor's control gear and arranged to adjust the engine speed to balance the air demand. Illustrated is the R4504 Hydrovane compressor developed from the earlier model and dating from the Sixties.
(150: Binks-Bullows Ltd, 1962)

150

151 A Dutch Land Rover One Ten Hard Top converted for transport of window glass by fitting a full-length steel tube roofrack – for ladders and lathes – and two side racks for adequate weight distribution. The vehicle is fitted with heavy-duty rear springs, and in the back there is a well-equipped tool set for setting new and replacing broken windows.
(151, 152: R de Roos, 1993)

153

Hydraulically-operated soil sampling rig, manufactured and fitted to a Series IIA Land Rover 109in Forward Control Chassis Cab by Boyles Brothers Drilling Co Ltd. This drilling equipment (type BBS 10) consists of a hydraulically-driven auger coring machine – its pump being driven from the Land Rover's centre pto – with a 12ft pull drilling mast to facilitate the use of 10ft rods. These vehicles were manufactured according to the requirements of the Road Research Laboratories, who used them during the Sixties. Boyles Brothers ceased trading at the end of the Seventies.
(153: British Drilling Association Ltd, 1968)

152

154

A light hydraulic work platform based on the Defender One Ten High Capacity Pick Up, fitted by the Dutch specialist Land Rover dealer Automobielcentrale De Uiver bv. It was supplied for use by the Shell petrochemical industry in Latin America in 1992.
(154: Automobielcentrale De Uiver bv, 1993)

Expeditions...

A Series III Land Rover 109in Station Wagon-based outside-broadcast sound vehicle equipped with antenna mast and air conditioning system, supplied to the Nigerian RTK (Radio Television Kaduna) by Pye TVT Ltd in 1975. During the Seventies, Pye TVT offered a wide range of these vehicles, equipped with radio transmitter/receiver units according to requirements, to broadcasting corporations worldwide, including the BBC. They were equipped with an audio mixing system, an operator set and a telescopic antenna mast to record programme material (spot recording) as well as to transmit live signals by microwave.
(155: Pye TVT Ltd, 1975)

155

157

156

For expeditions such as the British Trans-Americas and Camel Trophy, heavy-duty workboats are used as a bridging system for crossing rivers in remote and uninhabited areas where vehicle-deepwading is impossible. The workboats are fitted with an aluminium ladder/ramp system to ensure the right positioning and distribution of vehicle weight, and for its easy roll-on/roll-off manoeuvring. Two outboard engines are mounted to this, producing up to 28 knots. Each workboat has a single inflatable keel, five separate buoyancy chambers, sturdy fittings and reinforcement. This bridging system was designed by Avon Inflatables Ltd.
(156, 157, 158: Avon Inflatables Ltd, 1992)

158

159

The Camel Trophy Discovery conversion, driven by the winning Dutch team in the USSR in 1990. Up to now Series III short and long-wheelbase, Ninety and One Ten Station Wagons, Defender 130s and Range Rovers have been converted to the Camel Trophy specification by Special Vehicles. This conversion is fitted with extra bonnet lockers, a bullbar, a 3.5-ton electric winch, raised air intake, underbody and fuel-tank protection plates, extra front-mounted lighting, a rear door-mounted ladder, a roof-rack with sand channels, axe, railroad pick, shovel, jerrycans, Pelican suitcases, rollcage, Terratrip computer, hand-operated spotlight and baggage mesh.
(159: R de Roos, 1991)

160

A pair of One Ten-based safari expedition vehicles, which were used for a three-year round-the-world trip, and built and fitted out to the customer's special request by E V Engineering Ltd.
(160, 161: E V Engineering Ltd, 1992)

161

A Land Rover 127-based safari support vehicle, designed to carry stores, tents and so forth. It is also fitted for use as a mobile kitchen. E V Engineering Ltd also manufacture many other special-body Land Rover conversions including site-inspection vehicles for VIPs, field office and laboratory vehicles, minibuses and luxury motorhomes. (162: E V Engineering Ltd, 1992)

162

164

Land Rover High Capacity Pick Up fitted with a general purpose machine gun and raised gunner's seat – one of an order of six for the SWAT control team for operating in Africa. This conversion was built by Glenfrome Engineering Ltd.
(163, 164: Bristol & West Photography, 1992)

163

Fire-fighting...

165

16

167

The first fire engine conversion, based on an 80in pre-production Land Rover. During the early Fifties such Land Rover-based conversions were built by The Rover Company Ltd rather than by a specialist manufacturer. Fitted to the short-wheelbase Series I was a self-priming Pegson water pump which had an output of 200 gallons per minute and was driven by the rear power take-off. This pump supplied two full-size delivery hoses – contained in stowage lockers over both rear wheelarches – and, if necessary, the first-aid hose on the hose reel from an inboard 40-gallon water tank. Further fire-fighting equipment was supplied by Ajax. This compact fire engine – with optional metal cab – was particularly useful for fire-fighting in city centres, forest areas and factory sites. The drawing illustrates a Series II 88in petrol model-based fire tender with the later 1958 fitment of a KSB water pump.
(165, 166: Pegson Ltd, 1948; 167: Sieberg bv, 1958)

168

A mid-Sixties HCB-Angus Firefly Chassis Cab-based fire engine, the front of which is fitted with a Coventry Climax pump, driven by the front end of the engine's crankshaft, and capable of producing 2,270 litres per minute. This conversion is equipped with a first-aid water tank and hose on reel, 24ft light-alloy ladder, chromium-plated electric siren, suction hose, chromium-plated spotlight, a fire extinguisher mounted to the front of the vehicle, 10 lengths of delivery hose and foam cans. Only export models were fitted with a tropical roof. HCB-Angus Ltd manufacture a wide range of fire-fighting, rescue and emergency vehicle conversions based on Series IIA, IIB Forward Control, III and One Ten Land Rovers for use by civil and industrial and foresty brigades.
(168: British Motor Industry Heritage Trust/Rover Group, 1966)

69

The Angus Firestrike is a light fire appliance based on the Land Rover 109in Series III. Compact and versatile for use in light industrial fire-risk conditions, it is fitted with a 450-litre first-aid water tank, a pto-driven water pump, hose reel, generous locker accommodation and suction and delivery hoses. Two platform steps which may be used for additional standing crew are provided at the rear of the vehicle.
(169: British Motor Industry Heritage Trust/Rover Group, 1978; 170: HCB-Angus Ltd, 1978)

170

The V8 petrol Land Rover One Ten gives the light fire appliance a rapid response capability for rescue work and other emergency situations in rough terrain or inaccessible rural locations and industrial areas. It carries a water tank, rear-mounted pto-driven water pump for a 2,200-litre per minute output, first-aid hose reel and three-piece ladder, and is available in open Pick Up or fully enclosed rear locker-moduled body styles.
(171: HCB Angus Ltd, 1990)

171

172

173

Carmichael International Ltd, founded in 1849, and manufacturer of custom-built four and six-wheel-drive Land Rover Series II, IIA, III and One Ten chassis-based fire-fighting vehicles for more than 30 years, became the largest UK producer in this field, its extensive range comprising all relevant military and civil airport applications, including these examples.
(172: The James L Taylor Collection, 1992; 173, 175, 176: Carmichael Ltd, 1993, 1993, 1992; 174: Land Rover Ltd, 1978; 177, 178: Nick Dimbleby, 1993)

174

175

176

177

178

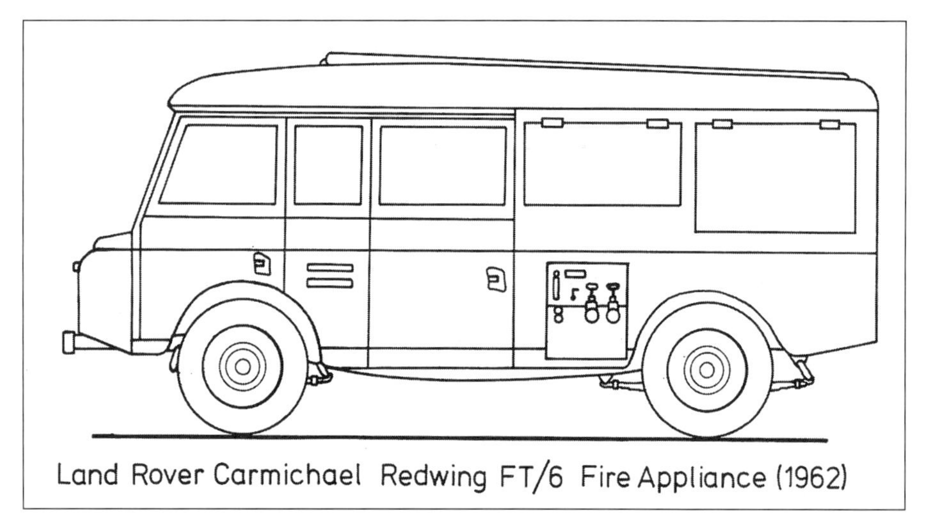

179

One of Carmichael's more ingenious bodies – manufactured during the early Sixties – was a Forward Control fire engine conversion known as the Carmichael Redwing FT/6, based on the Land Rover 109in chassisframe. This compact, manoeuvrable, fast and economical fire engine could carry a crew of four firemen, a two-piece ladder, suction hoses on hose reel, nozzles and a first-aid water tank. It was fitted with a pto-driven Coventry Climax water pump of 300 gallons per minute output.
(179: R de Roos, 1993)

180

This six-wheel-drive Carmichael fire engine was fitted with a roof-mounted Stem-Lite – produced by Zumro bv – which provided a blue-xenon flashing beacon and/or floodlights of 2.5m elevation.
(180: Carmichael Ltd, 1993)

181

Highly-equipped rescue tender, from Angloco Ltd, fitted with electric generator, extending floodlight mast with three 500W floodlights, winch and a wide range of rescue tools and equipment. Angloco Ltd is an independent company established in 1965, whose customers include airports, oil refineries, municipal fire brigades, US Department of Defence, UK Ministry of Defence, Civil Aviation authorities and public utilities including the Forestry Commission, the National Coal Board and British Telecom. Its full range of products includes air-crash rescue and rapid intervention vehicles, all-terrain fire-fighting vehicles, breathing apparatus appliances and control units.
(181: Angloco Ltd, 1978)

A water tender with on/off road capability. This means it can cover a wide range of accidents, offers rapid response and is highly manoeuvrable, allowing it access to areas where larger appliances would be restricted.
(182, 183, 184: Angloco Ltd, 1978)

182

183

184

A Hard Top Land Rover One Ten fire engine with one aluminium alloy roller shutter on each side and a central rear door which can be removed to provide an open pump bay. Its GVW is 3,050kg when fitted with a 100-gallon steel water tank and a light-alloy rear-mounted pump – for 300gpm output – an extension ladder, hose reel and suction hoses.
(185: Angloco Ltd, 1986)

185

186

187

Land Rover Ninety Pick Up fire command vehicle/dry powder unit. Charged with 250kg of BC powder, it has a separate nitrogen expellent and two all-aluminium alloy lockers, each with a hinged side flap. Internal illumination is fitted.
(186: Angloco Ltd, 1986)

Rear of a Land Rover One Ten High Capacity Pick Up-based fire engine by Kronenburg, one of Holland's few manufacturers of fire engines based on Land Rovers.
(187: Kronenburg bv, 1989)

188

Series III Land Rover 109in chassis-based fire-fighting vehicle, painted in a red-and-white colour scheme and designed by Kronenburg.
(188: Kronenburg bv, 1986)

189

190

Series III Land Rover 109in Pick Up fire engine conversion, manufactured in the Seventies by Dennis Eagle Ltd. It is fitted with a separate portable fire-pump unit, powered by a light petrol engine, for use in places inaccessible to a fire engine with a pto-driven pump.
(189, 190: Dennis Eagle Ltd, 1977)

191

192

A Pilcher-Greene Branbridge Mk5 fire appliance based on the Land Rover One Ten, providing seating for the driver and four crewmen, and fitted with a pto-driven Godiva water pump of 500gpm output at 6.8 bar, two suction hoses, a 19mm bore first-aid hose (55m), 315-litre water tank and 7.75m rope-operated ladder. Pilcher-Greene Ltd have been involved with wide-ranging Land Rover conversions for over 30 years including ambulances, tour and open review vehicles, handicraft vehicles, motor caravans, dispensary and clinic vehicles, and military staff and aircraft service vehicles. (191, 192: Pilcher-Greene Ltd, 1991)

One Ten High Capacity Pick Up-based Branbridge Mk4 fire appliance. This conversion has a 500-litre first-aid water tank, large transverse locker tilted above the tank, suction and delivery hoses, ladder and Godiva water pump. Its GVW is approximately 2,065kg.
(193: Pilcher-Greene Ltd, 1991)

193

194

A comprehensive range of Land Rover-based fire appliances is manufactured by V F Specialist Vehicles Ltd, of Marsden, Huddersfield for 4x4 or 6x6 chassis, and may be equipped with a 500gpm water pump, 66 to 235-gallon water tanks and foam tanks of 14 to 200 gallons capacity, in full Hard Top, semi Pick Up or Pick Up body styles. These vehicles are particularly well-suited to small airfields, chemical works and factories.
(194: V F Specialist Vehicles Ltd, 1992)

195

Defender-based six-wheel-drive, high-roof fire engine conversion with large single-axle water-tanker trailer attached. A special light-alloy extension ladder is fitted on the roof. Both the fire engine and the trailer are part of the range manufactured by Reynolds Boughton Ltd.
(195: Reynolds Boughton Ltd, 1990)

Generators and air conditioning...

During the Seventies Allam Generators Ltd manufactured Series III Land Rover diesel generator conversions to provide a substantial power supply suitable for the most unfavourable conditions, including extremes of temperature and altitude, and in remote off-road situations. This conversion consists of a 220/110-volt, 7.5kVA generator, driven via three V-belts from the gearbox power take-off. The generator is installed between the front seats so that the vehicle's rear carrying capacity is not reduced. The instrument panel for this conversion includes an ammeter, voltmeter, frequency meter, hours-run meter, and earth leak and overload circuit breakers. All the equipment is fully protected against mechanical stresses and shocks caused by rough off-road driving and moving loads in the back of the vehicle. Full protection against electrical overloading is standard. At continuous full load the consumption of diesel fuel by the generator at an engine speed of 2,500rpm (generator speed 3,000rpm) is 4.5 litres per hour.
(196: Allam Generators Ltd, 1976)

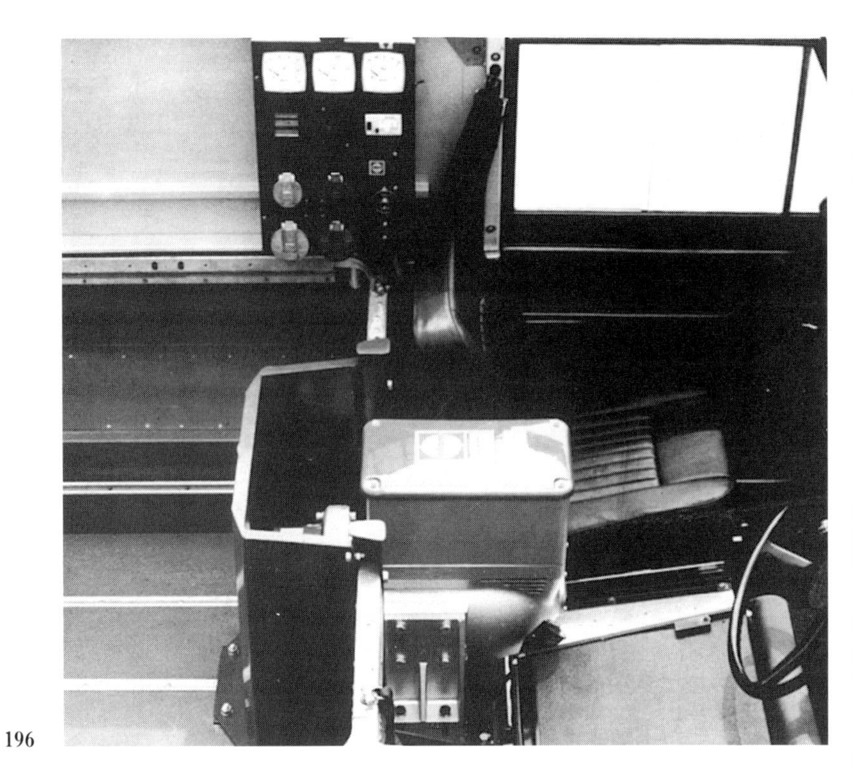

196

197

198

One of the first types of air conditioning system suitable for Land Rover installation was manufactured by Normalair Ltd and became available in 1964 for the normal-control petrol model only; installation for forward-control and diesel models was at that time still under investigation. This particular system consisted of a twin-cylinder refrigerant compressor – belt-driven from the engine crankshaft – an electro-magnetic clutch, which was incorporated into the compressor pulley, a condenser – installed in front of the engine's radiator – a liquid receiver, which acted as a refrigerant reservoir fitted with a sight-glass to indicate the correct working of the system, and an evaporator unit, comprising inlet and two outlet grilles fitted above the vehicle's instrument panel, the former of which circulated air from the vehicle's interior by two motor-driven fans. During this circulation, the heat exchanger matrix of the evaporative unit absorbed the heat from the interior air. All these components were installed within the engine compartment, except for the evaporator unit, which was installed partly here and partly in front of the vent-flaps under the windscreen. This system was completely sealed, and filled with dichloridifloromethane gas. A running vehicle engine and closed vehicle doors and windows were necessary for good system operation, and to gain the optimum from it the grille-mounted vehicle headlights were moved to the front wings, providing a maximum airflow through the condenser and engine radiator.
(197, 198: Normalair-Garrett Ltd, 1964)

199

200

201

Many types of Land Rover conversions and standard Land Rover Hard Tops/Station Wagons exported to hot climates are fitted with an external roof-mounted air conditioning system manufactured by Hubbard Transport Refrigeration Ltd. This 4.1kW capacity system – sufficient to suit every Land Rover body configuration – is designed to suit individual customer requirements. The complete 46kg system, including the cooling condenser, is mounted on the vehicle roof instead of in front of the engine radiator/oil cooler. This ensures a full cooling airflow through the engine radiator – important in hot climates – and also allows for good accessibility for maintenance and servicing. The adjustable thermostat for temperature control, a three-speed fan-motor control switch for air control and two adjustable air louvres, are positioned in the vehicle's roof, so no dashboard space is needed. Heavier, dashboard-mounted models of Hubbard air conditioning systems, with a higher capacity, specially designed for the Land Rover Series III and Ninety/One Ten, are also available. The Land Rover ambulance illustrated in (201) is manufactured by Freight Bonallack Ltd.
(199, 200, 201, 202, 203: Hubbard Transport Refrigeration Ltd, 1980)

202

203

204

205

Defender One Ten-based Supersilent fully integrated generator unit with a power output of 25kW and manufactured by A Smith, Gt Bentley, Ltd. The generator is not powered via the Land Rover's power take-off, but by a second, Volkswagen, diesel engine installed in the rear with the generator. The generating set itself is installed within a purpose-built, sophisticated and very compact lightweight acoustic enclosure, of double-walled construction, using the cavities for air inlet, outlet and exhaust attenuation ducts. The Supersilent's noise level is reduced to 58dB(A) at 3m from the vehicle during full power supply. Access is via hinged doors to each side, and the units are offered with or without stowage areas. The vehicle roof forms a strong working platform on which lashing rails are fitted. The available voltages from the vehicle are 110–220 volts and 120–240 volts single-phase, and 220/380 volts and 240/415 volts 3-phase; the available AC frequencies are 50Hz and 60Hz. The control panel is positioned on the left side of the enclosure, and is equipped with all necessary controls and instruments including a stop/start facility, gauges for engine oil pressure, water temperature, engine running-hour counter, fuel contents and an ammeter. Warning lamps are also provided for engine oil pressure and water temperature, low fuel contents, battery charge failure and engine overspeed, and the unit will automatically shut down if something goes wrong, such as a serious drop in oil pressure or excessive increase in water temperature. These generator units are manufactured according to customer requirements and are used by broadcast corporations including the BBC. (204, 205: A Smith, Gt Bentley, Ltd, 1994)

Leisure activities...

A Pilcher-Greene caravan conversion based on the Series IIA Land Rover 109in, with walking access on the roof and flush-mounted steps.
(206: Pilcher-Greene Ltd, 1968)

206

207

A Land Rover Hard Top converted to a mobile kitchen for the purpose of supplying food and drinks to visitors of hunting and shooting parties on a country estate.
(207: E V Engineering Ltd, 1992)

208

The Land Rover 127 Barouche, designed for activities such as fishing, shooting and attending equestrian events, and fitted with refrigerator, oven/grill, water system and an elevating roof to allow standing room for cooking and outside observation. It has no rear door.
(208: E V Engineering Ltd, 1992)

209

210

A Luton-type coachbuilt unit providing sleeping accommodation for four and incorporating a shower room, kitchen and lounge area, all of it air conditioned.
(209, 210: E V Engineering Ltd, 1992)

211

One of the first British manufacturers of Land Rover-based motor caravans was Martin Walter Ltd, since trading as Dormobile Ltd. In 1962 they made their first Land Rover motor caravan conversion based on the 109in Series II Station Wagon, which was later followed by similar conversions to the IIA and III Station Wagons. This particular conversion was known as the Dormobile Caravan and offered supreme versatility for comfortable touring by four adults. The patented glassfibre elevating-roof construction provided ample headroom, air space and ventilation. The motor caravan was equipped with cooking, water supply, storage and washing facilities, and made an ideal vehicle for long-distance holidays.
(211: Dormobile Ltd, 1967)

212

Carawagon Ltd also converted long-wheelbase Land Rovers into raised-roof motor caravans. The company was taken over by Woodflair Ltd, who then ceased trading in 1986. The Land Rover Woodflair full-length raised-roof conversions provided accommodation for two to four people and could be fitted with hardwood doors, water tanks, a cooker, gas bottles, a sink, custom-built furniture, settees/beds, thermostatically-controlled heater/air conditioning, an engine-heated warm-water preparer and polished wood-panel trimmed interior, all fitted according to customer requirements. Some of these units were supplied to the British Army, who adapted them for use as a light and compact field command post for combat, communication and intelligence operations. It had sleeping provision for two people, and its workspace could be extended by a canvas tent attached to the rear of the body. The vehicle was fitted with FM radio communication equipment, antennae, extra batteries, a writing table, roller blinds, a field telephone set, desk lamp, front-mounted roofrack and prick-board for detail maps.
(212: Land Rover Ltd, 1987)

213

Action Mobil, of Austria, manufactures demountable and fixed camper units for long-wheelbase Land Rovers. Each aluminium alloy unit is custom-built and may be fitted with a heater, gas bottles, general cupboard space, 110-litre water tank, washing and cooking combination, refrigerator, toilet, extra batteries and a wardrobe. These units are also manufactured for ambulance purposes.
(213, 214, 215: Action Mobil, Austria, 1993)

214

215

216

Autarkia Camperunits, of Holland, designed and developed a clever, high-quality camper-unit system which accommodates four people. It can be used with the Defender 130 Crew Cab Pick Up, for which Autarkia has developed a unique stainless steel quick-release system. This means that the original rear body section may be detached or removed from the chassisframe in a couple of minutes, making it into a Defender 130 Chassis Crew Cab, on which the camper unit can be fitted. The camper unit is fitted with high ground-jacks to lower it to the Defender chassisframe during attachment. This system allows for a more spacious and complete camper unit to be fitted, the removal of the original rear body section saving weight and thus allowing more freedom in designing the interior. The exterior length and width of this camper unit is 3.85m and 1.98m respectively and its floor length is 0.50m longer than that of the original Defender Crew Cab rear body section. The rigid, aerodynamic body consists of 34mm thick insulated sandwich panels. Standard fittings include a refrigerator, warm-water facility, pressurized water tank, electronic unit to transform the 12V DC battery to 240V, an extra 12V 75Ah battery, toilet, shower, double-insulated windows, hot-air heater, full interior lighting, a sink, cooker, cupboard, bedroom, wardrobe and gas bottles – but again these may be subject to customer requirements. The weight of this camper unit is approximately 680kg and it is also available for the Defender One Ten and 130 Chassis Cab.
(216: Autarkia Camperunits, 1993)

217

A mobile kitchen is one of the many applications for which the highly successful Quadtec box body can be used, the conversion being produced by Special Vehicles. Based on the Defender 130 Chassis Cab, it is fitted with large gas strut-assisted top-hinged hatches. Such a unit may be fitted with a raisable roof to give standing headroom, as well as a refrigerator, sink, cooking range, oven, gas bottles, fire extinguisher, fire blanket, water tank, warm-water facility, cupboards, a large dustbin, mini petrol generator for 220 volts and an electrical ventilation system.
(217: Land Rover Ltd, 1991)

218

Two Series III Land Rover 109in-based ice cream van conversions. These all-wheel-drive vehicles have the obvious advantage of being able to reach beaches and other off-road leisure spots during the summer peak-trade period, when a two-wheel-drive vehicle would find it hazardous.
(218: R Dekker, 1991)

Ninety and Discovery-based Amphy Rover Special Vehicle conversions used exclusively during large public events to promote Land Rover products. This type of amphibious conversion is fitted with a flotation kit – from Avon Inflatables Ltd – consisting of five airbags held by glassfibre shrouds, mounted to the vehicle's chassis-frame by five steel-tube frames. Both the Ninety and the Discovery are fitted with Avon Tredlite low-pressure tyres which enable this conversion to drive on very soft ground. Both vehicles are fitted with a pto-driven 12in diameter propeller which gives enough propulsion for a speed of 4 knots. In 1959, RFD Ltd used similar techniques to convert some long-wheelbase Land Rovers to amphibious vehicles for the British Army. (219: Avon Inflatables Ltd, 1989; 220, 221, 222, 223: Harry Andriessen; Don van der Vaart Fotoproductions bv, 1993)

219

220

221

222

223

224

A high-quality car tent to fit the rear of the Land Rover Defender, Range Rover and Discovery, manufactured by Caranex. This one-piece tent is made of a specially woven nylon, and has only three poles, giving a floor area of 2.25 x 1.5m and 2m headroom. It is easily assembled in six minutes, and is ideal for more serious long-distance travelling.
(224: Caranex, 1992)

225

To reduce the high cost of purchasing factory-built Land Rover motor caravan conversions, many camper units are being built by individual enthusiasts, based on all types of Land Rover.
(225: R Dekker, 1993)

Another example of a home-built camper unit, based on the Land Rover One Ten. This has been extended to the 6x6 format and fitted with the rear section of a Marshall of Cambridge-built military ambulance body.
(226: Nick Dimbleby, 1993)

226

227

In 1957, one of the leading golfball manufacturers, Penfold Golf Ltd, developed a comprehensive test programme for their products to demonstrate the comparative properties of existing golf balls on their test grounds. To gain more publicity and allow club members to watch this test on their own golf course, the company developed a Robot Driver demonstration vehicle, based on a Land Rover 107in Series I. This showed the balls' accuracy of flight, the difference in trajectory between British and American-size balls and many other items of interest to golfers. The vehicle had two separate driving surfaces, operated by the Land Rover rear power take-off, from which balls could be struck either separately or simultaneously.
(227: R de Roos, 1993)

Drawing of a Series III Land Rover 88in Pick Up at the Dutch glider ground of Castricum, near Amsterdam. This was used to pull three 1km-long cables simultaneously from a very large, stationary drum winch, powered by a 170hp petrol engine and fitted with three separate cable drums, to the point where the gliders were waiting to be hooked on and pulled. When a glider was pulled high enough into the sky it dropped the cable, the top of which was provided with a small parachute to reduce its rate of fall. When the parachute landed, the cable would be completely rolled up around the cable drum. From that moment it was allowed to use the second cable to pull the next glider into the sky. When the three cables were used and rolled up onto the cable drums, the Land Rover arrived to hook on the three cable-ends and to pick up the small parachutes which were put in a specially-made pneumatically-operated transverse tipping tray. Further, the Land Rover was driving – dragging the three long cables behind it – to the point where the next gliders were waiting to be hooked on. When the Land Rover arrived at this glider waiting point, the tipping tray dropped the parachutes. The three cables were also dropped. The Land Rover was fitted with a back-mounted steel bar on which three hooks were welded, which could be rotated by a left-mounted pneumatic cylinder to drop the cables from the hooks. After that the Land Rover was driven away from the glider waiting point and the first glider could be hooked onto the cable.
(228: MWK, 1972)

228

During 1987 and 1988 a tracked, V8-engined Range Rover was used for ski-slope service in New South Wales, Australia. Four separate track units were bolted directly onto its wheelhubs, using only the five studs and nuts of each hub, after the normal road wheels were removed – similar to the tracked Land Rovers which were manufactured by J A Cuthbertson during the Sixties, but with a completely different suspension. Because of the third (central) differential, all of the vehicle's tracks were driven without axle shaft wind-up. Despite fitment of power-assisted steering, the turning circle was very poor, and quite often the steering assembly of the vehicle was damaged due to the excessive force which was required to turn the heavy front track units. This was not a successful project, therefore, and after a few years the expensive track units, which had been imported from the United States, were removed from the vehicle and eventually discarded.
(229: Rover Australia, 1987)

229

Mast vehicles...

230

A television surveillance Land Rover 109in V8, fitted with two masts. Clark Masts Teksam Ltd supply and fit telescopic, air-operated masts to many different vehicles, including Land Rovers and Range Rovers. They extend to a height of 21m and can support heavy headloads as well as delicate antennae. The applications for the masts are many and varied, and in addition to the broadcasting industry, include military work and usage on Camel Trophy support Land Rovers.
(230: Clark Masts Teksam Ltd, 1981)

231

Specially adapted for British Gas, this Land Rover One Ten, with its big 10m air-operated telescopic mast, was to be fitted out with electronics to serve as an emergency base station. Carrying at least one microwave dish, to be accompanied by VHF and UHF antennae, its main function would be direction finding: helping to overcome the problems of radio interference in the gas authority's operational communications system. The retractable and fully-guyable mast – operated by an onboard compressor – has a base tube diameter of 8in and can support headloads up to 140kg. To provide additional stability the vehicle is fitted with stabilizing jacks.
(231: Clark Masts Teksam Ltd, 1990)

232

Air-operated telescopic mast mounted on a Land Rover One Ten for the Welsh Water Authority, totalling 15m in height, and used to support the folded dipole communications antennae and two battery-operated sodium floodlights employed for night-time emergency work. When the vehicle is mobile the mast is safely nested and secured in its roof trolley.
(232: Clark Masts Teksam Ltd, 1989)

234

233

Close-up of the special adjustable mobile mounting system for the air-operated telescopic mast. Prior to extension for in-the-field communications, the mast may be set vertically by an in-built spirit level. The adjustment to the lower mounting bracket ensures that the mast can be set upright even when the vehicle is parked on uneven ground.
(233: Clark Masts Teksam Ltd, 1989)

A police Land Rover One Ten Station Wagon fitted with a lighting mast.
(234, 235: Clark Masts Teksam Ltd, 1987)

235

236

A 20m Clark mast mounted on an ex-British Army Series III Land Rover 109in Pick Up. This conversion is used by a Dutch photographer as a photo-camera mast for customer-request aerial view photographs. The vehicle is fitted with a pto-driven 10 bar air compressor and 60-litre buffer tank. A 4 bar pneumatic cylinder is also fitted to erect the mast from the travelling position. Only 1.5 bar is needed to extend the pneumatic telescopic mast to its full height. A video camera is mounted parallel to the photo camera, allowing the latter to be pointed at the correct angle, and a vehicle-mounted monitor shows the operator the selected area that has to be photographed. The photo-camera/video-camera unit is mounted on a servo-assisted frame so that it can be tilted and rotated in any direction when raised to its full height.
(236, 237: H Uitdenbogerd, 1993)

237

In the late Sixties, Snell Aerial Photography developed a pneumatically-operated, seven-section telescopic mast to raise a photo-camera, film-camera and video-camera to 20m above ground level. The complete system, called Hi-Spy, could be fitted into the back of a short-wheelbase Land Rover Pick Up equipped with outrigger jacks in order to set the vehicle horizontally before raising the mast. Within a couple of minutes of arriving on site, photographs could be taken from full height.
(238: Snell Aerial Photography, 1978)

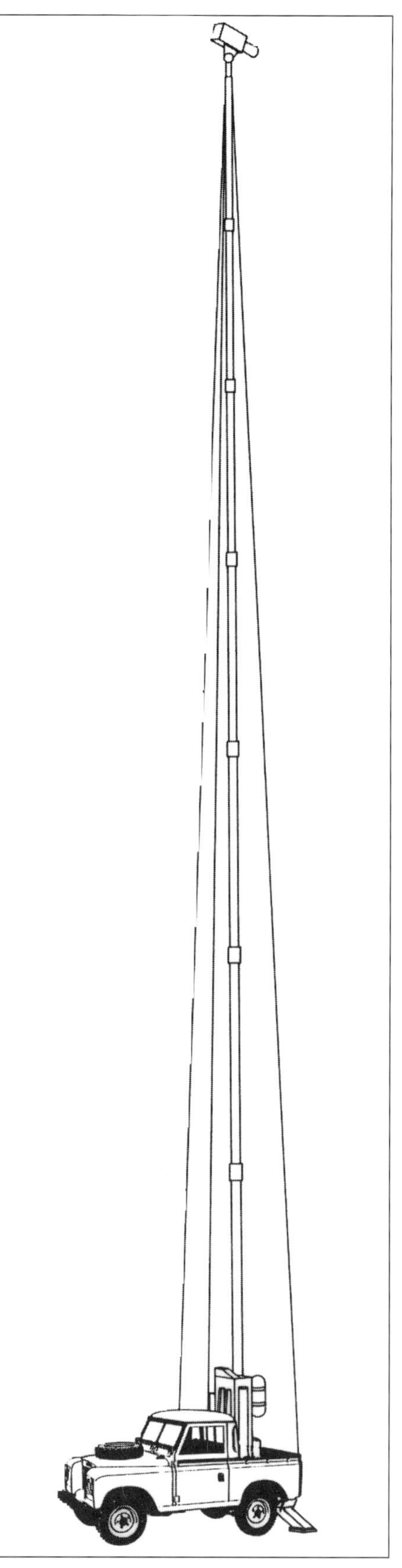

238

Medical transport...

239

A white-painted shockproof X-ray unit, sold to a Middle East customer. This Series III Land Rover 109in Station Wagon-based X-ray unit, with attached generator trailer for supplying the electric power, is intended to provide full X-ray treatment facilities to those living in remote areas, and was manufactured and supplied by Angloco Ltd.
(239, 240: Angloco Ltd, 1978)

240

241

242

Macclesfield Motor Bodies (UK) Ltd offers a conversion for the Defender Ninety or One Ten for the transportation of the wheelchair-bound. The conversion entailed replacing the standard rear door by new double doors, fitting a hydraulic wheelchair-lift and raising the original roof.
(241, 242: Macclesfield Motor Bodies <UK> Ltd, 1990)

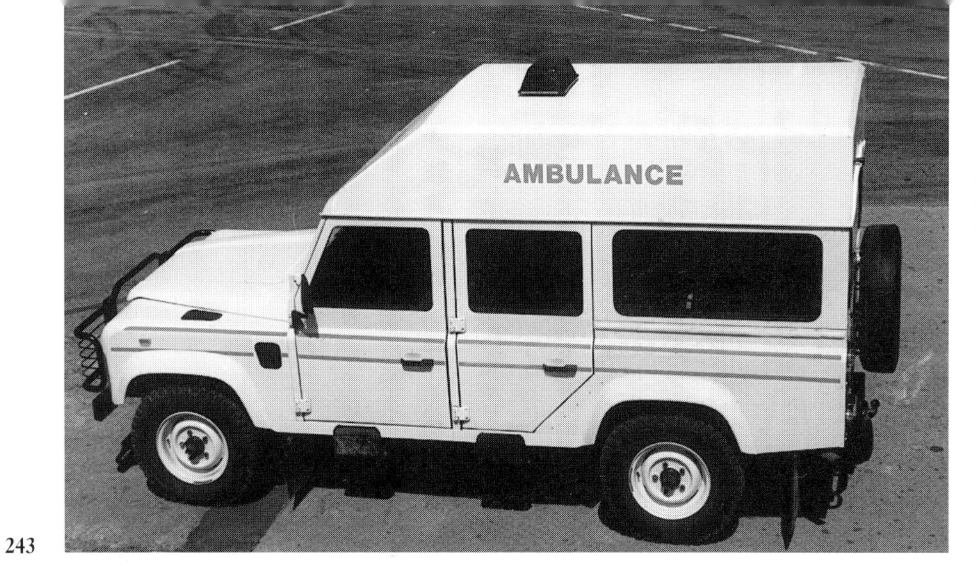

243

A Land Rover One Ten basic ambulance conversion based on the same body design as that shown in (83) and manufactured by Macclesfield Motor Bodies (UK) Ltd.
(243: Macclesfield Motor Bodies <UK> Ltd, 1990)

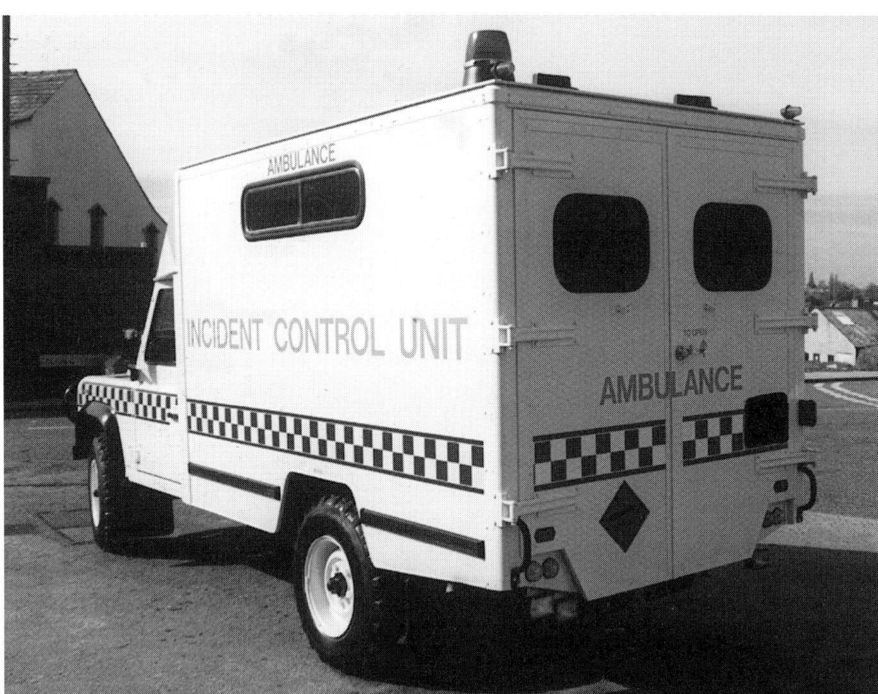

244

Many situations – at airfields, outdoor pop festivals, motorcross events, army locations and mountain/undeveloped areas – require rugged, powerful and durable four-wheel-drive all-terrain ambulances. Using Defender 110 and 130s as the basic vehicles, Macclesfield Motor Bodies (UK) Ltd make spacious and well-equipped conversions for medical authorities and governments worldwide. Patient safety and comfort have been the priorities in the design of these ambulances, which are available in five different versions to accommodate stretchers, life-support systems, ancillary equipment and medical attendants. The ambulance body is insulated and can be fitted with such equipment as a fresh-air intake facility, rotating/flashing beacon, an emergency two-tone horn, electric fans, an attendant's seat, infusion bottle clips on a sliding rail, comprehensive first-aid kit, personal oxygen system, etonox analgesic set, light-rescue tool kit, BCF fire extinguisher, radio communication equipment, emergency maternity pack, patient's carrying chair and two extra batteries. This company also manufactures Land Rover-based mobile clinics and cadaver units.
(244, 245, 246: Macclesfield Motor Bodies <UK> Ltd, 1992)

245

246

247

A Series III Land Rover 109in-based ambulance supplied to a Dutch customer and manufactured by Pilcher-Greene Ltd, in 1977.
(247: Fotografie Lex Klimbie, 1977)

248

Formed in 1925 to deal with the ever-increasing demand for ambulances and specialized medical vehicles, Pilcher-Greene Ltd was a leading pioneer in developing and supplying these conversions for Land Rovers, this early example being based on the Series I 107in Pick Up.
(248: Pilcher-Greene Ltd, 1957)

249

The range of vehicles extend from the smallest country ambulance to the largest operating theatre, and they are designed to meet the needs of physicians, surgeons and technicians working in all sorts of climate and terrain. Since the Fifties the company has been manufacturing medical conversions for Series I, II, IIA, III and One Ten Land Rovers, those for the One Ten being available in five body types, designated B, F, S, V and X. Shown here is a Type F, one of the Series 8404 low-cost ambulances, which was available in four different versions.
(249: Pilcher-Greene Ltd, 1986)

250

Type B, also from the Series 8404 and available in two different versions.
(250: Pilcher-Greene Ltd, 1990)

251

The Type S Series 8303 ambulance, available in five different interior styles. It has a wider rear body consisting of a welded mild steel framework with moulded GRP or aluminium panels, and is fitted with the latest ambulance equipment. The sides and roof of the robust light-alloy and GRP body structure are insulated with glassfibre wool, and this vehicle is air conditioned.
(251: Pilcher-Greene Ltd, 1989)

252

Body types V and X were Series 8505 dispensary/clinics. Their exteriors were similar to that of the B and F, and had as standard a three-sided 6ft canvas extension complete with guy ropes, which fitted into a channel in the rear, providing a sheltered waiting or treatment area. Shown here is a Land Rover-based Pilcher-Greene mobile clinic conversion, which was supplied to an Indian customer.
(252: Pilcher-Greene Ltd, 1982)

253

A Land Rover One Ten-based ambulance conversion, manufactured by Herbert Lomas Ltd, who for many years manufactured a wide variety of ambulance bodies for the 109in (Series IIA and III), One Ten and 127in Land Rovers.
(253: Nick Dimbleby, 1988)

Military use...

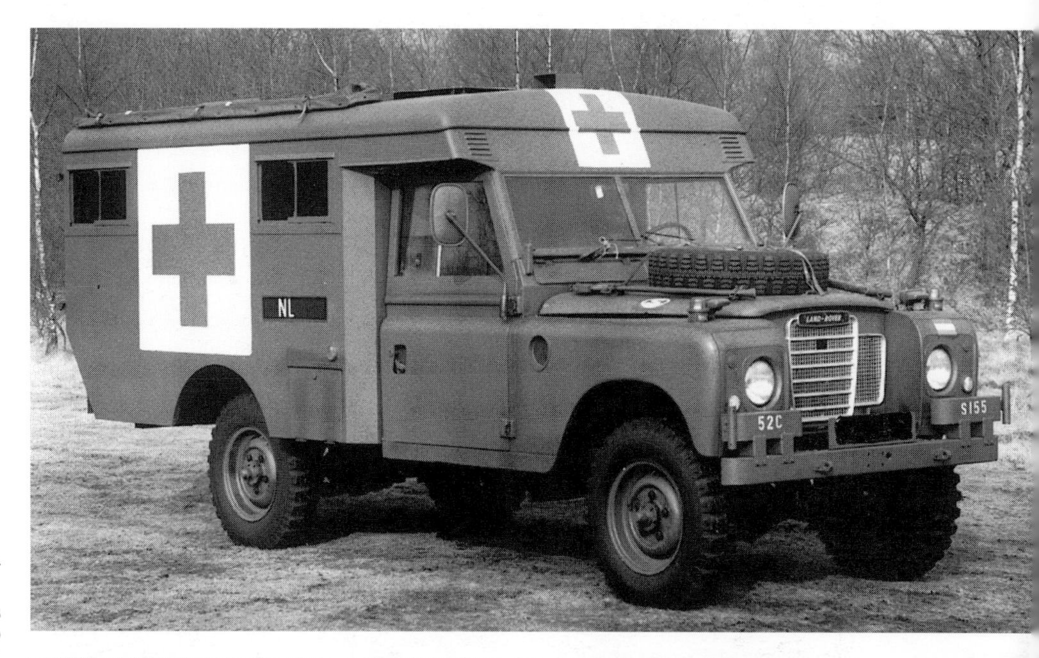

254

The military ambulance conversion has been sold in enormous numbers to NATO armed forces. Series II, IIA and III Land Rover 109in were used as the basic vehicles for this type of military ambulance introduced in 1960 and produced until 1982 by Mickleover Transport Company Ltd and Marshall of Cambridge (Engineering) Ltd. It provided accommodation for four stretcher cases and one medical attendant or six seated patients. The thermally insulated rear body was framed in aluminium and panelled in aluminium and hardboard, and both the cab and rear body were heated and ventilated. This ambulance was designed to provide casualty evacuation from forward areas. Because of the high centre of gravity and a maximum weight of 2.67 tons, which resulted in considerable body roll when cornering and inferior cross-country stability, the vehicle was fitted with an improved suspension and two anti-sway bars. The fixed front seats had to be moved some inches to the front to provide sufficient floor length in the ambulance compartment.
(254, 255: Ministerie van Defensie Directie Voorlichting, 1991; 256, 257: R de Roos, 1989)

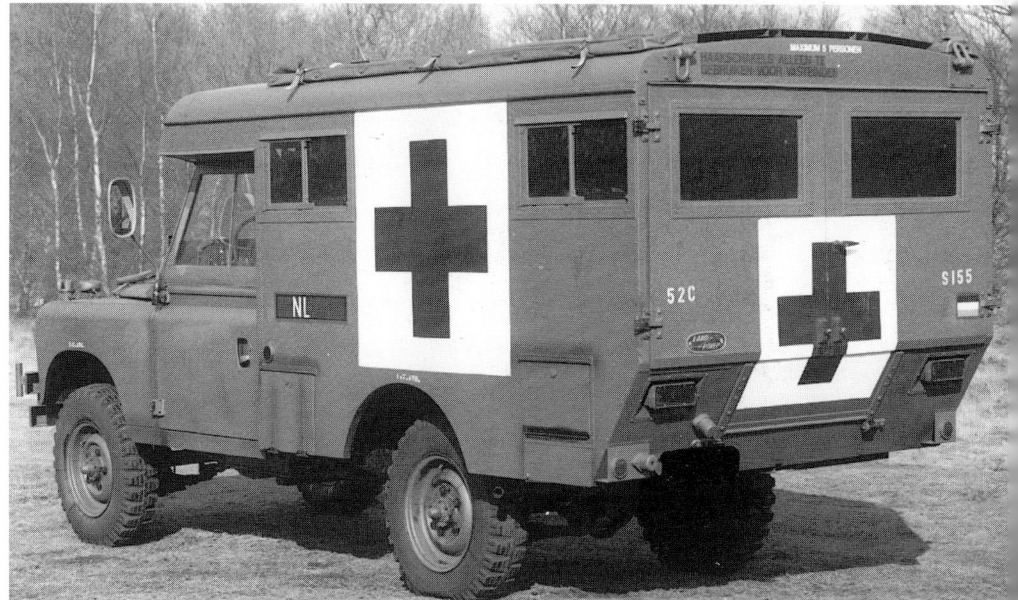

255

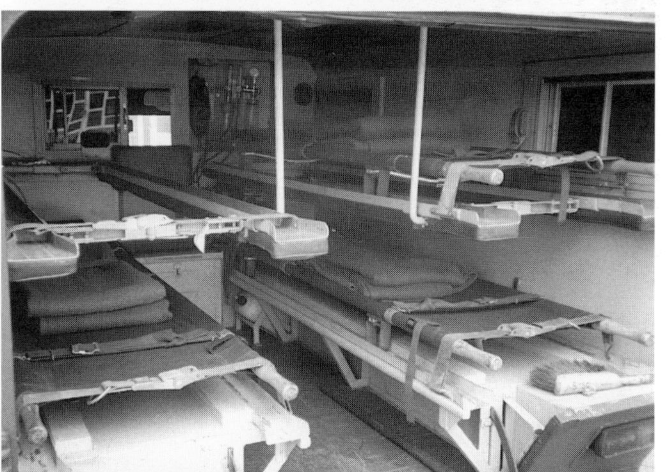

257

A Defender 130-based military ambulance in the service of the Royal Air Force and produced in large numbers by Marshall of Cambridge (Engineering) Ltd for the British armed forces during the early Nineties. It resembles the military ambulance shown in (267) manufactured by Macclesfield Motor Bodies (UK) Ltd, but note the subtle differences in the bodies.
(258: Marshall of Cambridge <Engineering> Ltd, 1993)

258

A Land Rover 130-based ambulance built by Marshalls for the armed forces of Oman, fitted with a Hubbard Engineering air conditioning system.
(259: Marshall of Cambridge <Engineering> Ltd, 1993)

259

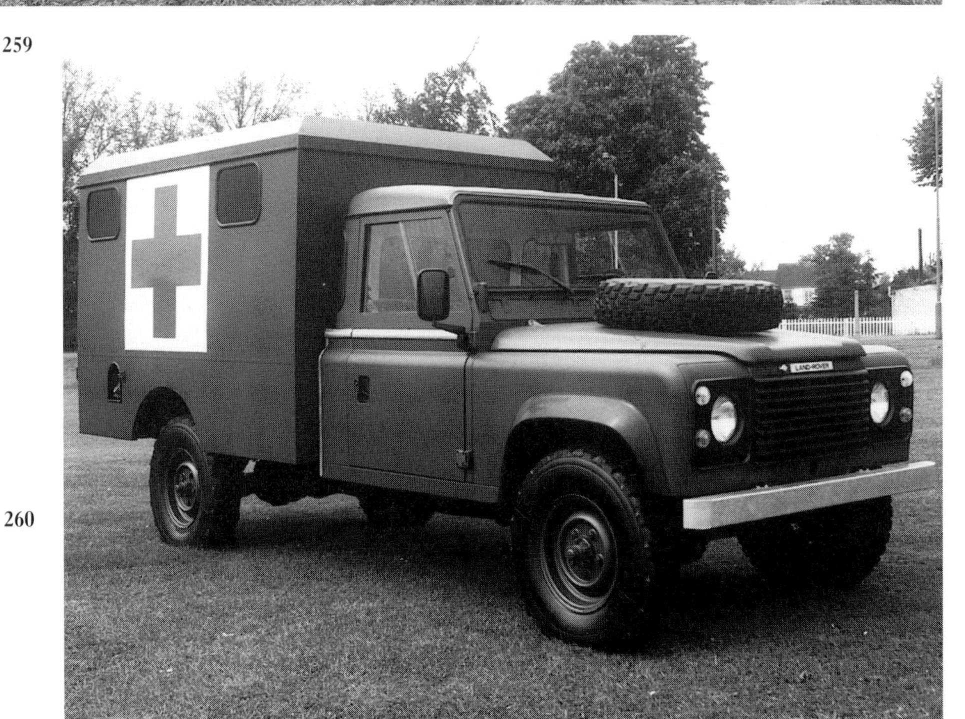

A box body-designed ambulance based on the Land Rover Chassis Cab, produced by Marshall of Cambridge (Engineering) Ltd.
(260: Marshall of Cambridge <Engineering> Ltd, 1993)

260

Another well-known military ambulance body, also manufactured by Marshalls, and fitted to the Series III Land Rover 101in Forward Control 1 Ton chassis and supplied in large numbers for the British Army during the Seventies.
(261: Marshall of Cambridge <Engineering> Ltd, 1977)

261

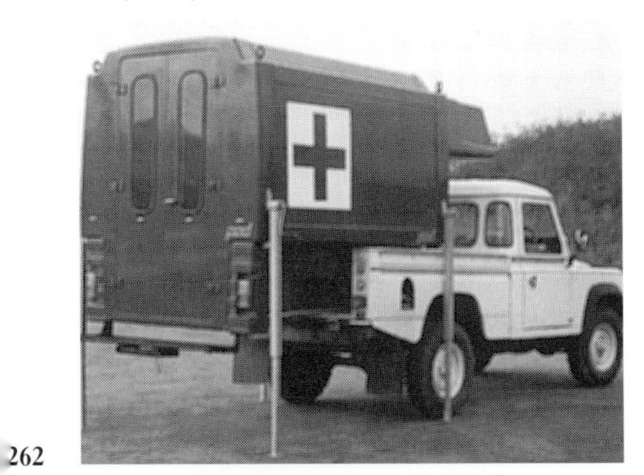

262

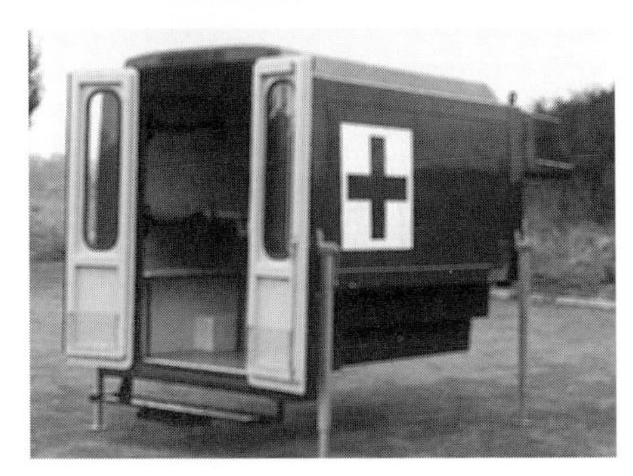

263

264

Fully air portable GRP demountable military ambulance unit, manufactured by Shanning POD Ltd. It may be fitted to the Series III Land Rover 109in Pick Up/Soft Top, One Ten Pick Up/Soft Top and One Ten High Capacity Pick Up/Soft Top. This conversion is fitted with four jacks – one on each corner – which allow it to be lifted from the vehicle. When the vehicle is driven away, the unit can be lowered until the folding rear steps reach the ground. Two side-hinged rear doors are fitted, providing easy access. The unit can be supplied with all the equipment necessary for a military ambulance, including seats, stretchers, roof-mounted air conditioning, an electronic unit for transforming the vehicle's 12 volts DC to 240 volts AC, full interior lighting, a water tank, small sink, first-aid kit and specialist medical equipment. The roof of the unit is fitted with special lifting eyes for shipping and lifting by helicopter.
(262, 263, 264, 265, 266: Shanning POD Ltd, 1992)

265

266

267

New Land Rover 130-based box body ambulances, manufactured by Macclesfield Motor Bodies (UK) Ltd, of which that illustrated in (267) was designed by Locomotors Ltd and has been sold in large numbers to the British armed forces during the early Nineties. Note the body differences when compared with the military ambulance shown in (258).
(267, 268: Nick Dimbleby, 1993; 269: Land Rover Ltd, 1993)

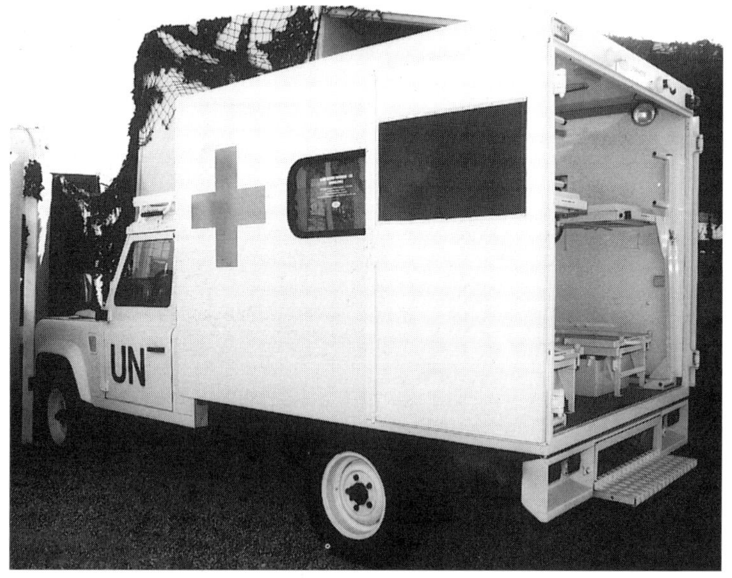

268

269

270

A purpose-built mobile workshop box body fitted to the rear of a Land Rover One Ten Chassis Cab. A pneumatic mast is fitted to the roof, which can be raised and pumped-up on the special rear mountings, access to the roof being provided by the rear steps. The body is fitted with special racking to accommodate the varied assortment of electronic equipment that was required for this mobile workshop. This conversion was carried out by Glover Webb Ltd in close collaboration with the customer in 1989.
(270: Glover Webb Ltd, 1989)

The 127-based general service box body, which may perform many military roles including use as a command control communications and intelligence centre, an electronic warfare vehicle or a mobile workshop. In 1989, The Directorate of Material of the Royal Netherlands Army carried out rigorous trials of some 127s before augmenting the familiar Land Rover 88in Lightweight, 109in general service and ambulance vehicles, some 4,500 of which have served the Dutch Army between 1978 and 1993.
(271: Fotografie Lex Klimbie, 1990)

271

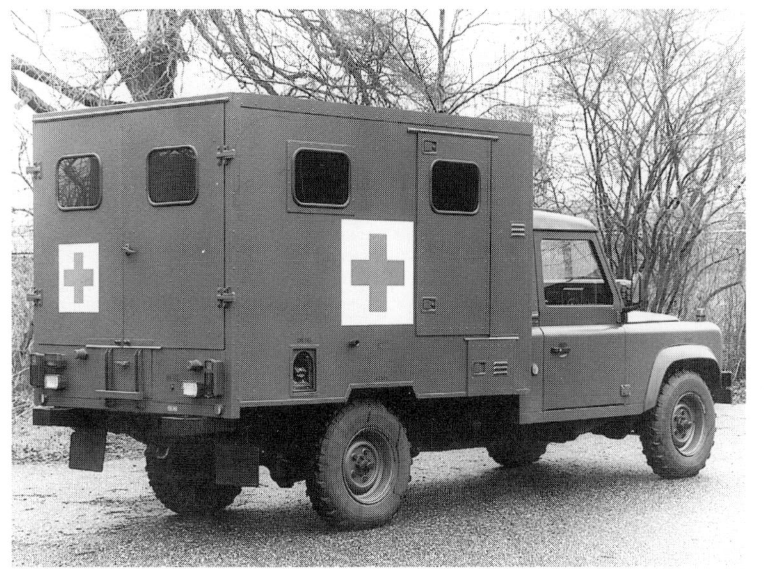

272

273

Examples of the box body of the Land Rover 127-based ambulance conversion, made by Marshall of Cambridge (Engineering) Ltd to the order of the Government and Military Operations Department of Land Rover Ltd and to the requirements of The Directorate of Material of the Royal Netherlands Army.
(272: Fotografie Lex Klimbie, 1990; 273: Marshall of Cambridge <Engineering> Ltd, 1990)

274

A light armoured Land Rover manufactured by Glover Webb Ltd and built for forward-area commando and reconnaissance roles. This conversion is specially designed to give a high cross-country capability, high on-road speed, protection against light ammunition and inconspicuous operating capability.
(274: Glover Webb Ltd, 1992)

275

276

These reconnaissance vehicles are fully armoured by a monocoque hull of special steel in which 35mm laminated bulletproof glass windows are fitted to provide protection against 7.62mm NATO ball ammunition, blasts and grenade fragments. The body is fitted with two narrow side-hinged rear doors, and extra protection is given to the fuel tank.
(275, 276: Glover Webb Ltd, 1992)

277

278

Land Rover Hussar Armoured Personnel Carrier conversion, a product of Penman Engineering Ltd. The lightweight high hardness steel armoured hull of this 6x6-driven vehicle is a fully welded body unit for a capacity of 14 men including the driver. It is designed to give full protection against 7.62 high-velocity rifles and machine guns, and grenade fragments. The floor is an integral part of the monocoque hull, which is bolted to the chassisframe and provides good protection against mine fragments. This type of conversion is used for many military and police roles such as a security/safety vehicle, armoured personnel carrier, a command/communication vehicle, ambulance, EOD transport and command vehicle, a cash and valuables transport vehicle, narcotics unit, prisoner transfer vehicle, escort vehicle, and patrol and reconnaissance vehicle. It is also available in the shorter and lighter 4x4 format, where it is known as the Skirmisher and is based on the Land Rover One Ten chassisframe. The GVW of the Hussar and Skirmisher is 4.5 tons and 3.8 tons respectively. A wide and varied range of fittings is available, including a fully rotating turret, night-vision facilities, a self-sealing fuel tank, air conditioning, smoke-grenade launchers and run-flat tyres. Penman also manufacture a wide and varied Trekker range of unarmoured conversions, based on the Land Rover 6x6 chassisframe, including mobile lubrication units, remote-area (SAS-type) patrol vehicles, many types of fire engine, breakdown recovery vehicles, debris clearance vehicles, ambulances, rapid intervention vehicles and emergency/accident tenders.
(277: Penman Engineering Ltd, 1990; 278: Nick Dimbleby, 1993)

Conversion based on the Series IIA Land Rover 109in, in SAS service during the Sixties and Seventies. This vehicle provided more load space than the Series I-based SAS vehicle and was fitted with 9 x 15in sand tyres, front and rear-mounted 7.62mm general machine guns, long-range fuel tanks, tyre-stripping tools, a curfew light, theodolite for stellar navigation, a grenade pouch, water jerry cans, sand ladders, magnetic and sun compass, a chain, smoke-grenade launchers, petrol stove and pans, spotlight, camouflage nets, long-range Morse radio, ground-to-air radio, rifle holster, signal pistol, first-aid box, three fire extinguishers, a pick axe and a slave cable. It was known as the Pink Panther because of its pinkish body colour, similar to that of the desert when viewed from the air. The SAS (Special Air Service) have been using Land Rovers since the Fifties, The Fighting Vehicle Research and Development Establishment having developed and converted the standard military Land Rover to a Remote Area Patrol Vehicle according to SAS specifications. Series-production of these long-range desert patrol and reconnaissance vehicles was carried out by Marshall of Cambridge (Engineering) Ltd. Until 1968, the Series I Land Rover 86in was used as a basic vehicle for this conversion, fitted with two Vickers machine guns, one Sterling machine gun, radio equipment and many other items.
(279: The National Motor Museum, Beaulieu, 1969)

279

The SAS conversion based on a Series III Land Rover 109in, also carried out by Marshall of Cambridge (Engineering) Ltd.
(280: Marshall of Cambridge <Engineering> Ltd, 1976)

280

Land Rover One Ten High Capacity Pick Up SAS vehicle conversion, manufactured by Glover Webb Ltd to NATO quality standard AQAP-1, and used as a remote area patrol vehicle for secret operations on the eve of the 1991 Gulf War. During that action some light radar-guided Iraqi missiles were captured in order to find the right radio frequency on which this missile could be jammed. These air-portable vehicles have a three-man crew and are able to make long-range desert patrols and reconnaissances of up to 500 miles without refuelling or other logistic support. They are fitted with a variety of equipment including 7.62mm general-purpose machine guns, a 12.7mm machine gun, grenade launchers, full radio equipment, thermal imaging supported by onboard pure air charging, satellite navigation equipment, a magnetic compass, sun compass, first-aid kit, camouflage nets, pioneer equipment, fuel-tank protection, two spare wheels, sand ladders, extra fuel and water tanks, and a shovel and pick-axe. Its GVW is 3,050kg.
(281: Glover Webb Ltd, 1990)

281

282

Reconnaissance and security vehicles based on the Series III Land Rover 109in Soft Top, as used by the Royal Dutch Army and Air Force. The conversion work – fitting the gun mounting for a 7.62mm general-purpose machine gun – was carried out by the Dutch Army themselves.
(282: R de Roos, 1989; 283: Commando Tactische Luchtstrijdkrachten, 1989)

283

284

285

Both 109in and One Ten Land Rovers are converted by the Royal Dutch Army for the Royal Dutch Air Force to give mobile short air defence capability to air force bases. These Land Rover conversions are fitted with a weapon ring-mount on which a Browning 12.7mm machine gun is fitted.
(284, 285, 286: Photographic Department Leeuwarden Air Force Base, 1990; 287: Commando Tactische Luchtstrijdkrachten, 1989)

286

287

An air-portable rapid intervention vehicle known as the Special Operations Vehicle, based on the Defender One Ten. Manufactured by Marshalls and produced since 1992, the body has been heavily modified in order to perform special military operations. It is equipped with a heavy-duty rollcage in which a weapon ring-mount is fitted, capable of firing in any direction. This vehicle is suitable for use as a remote area patrol vehicle and provides seating for a six-man team. It can be fitted with satellite navigation equipment, a 40mm grenade launcher, 7.62mm general purpose machine guns, a canvas cover, winch, ammunition storage boxes, radio communication equipment, heavy-duty suspension, a special gunner's seat and even the ASP 30mm machine gun.
(288, 289: Nick Dimbleby, 1993)

288

289

The Multi Role Combat Vehicle (MRCV), designed to meet the growing demand for a highly mobile weapons platform, based on the Defender Ninety and first shown publicly at the British Army Equipment Exhibition of 1993. The MRCV has the versatility and adaptability to perform several specialized combat roles to the requirements of the rapid reaction forces and reconnaissance/ patrol units, thus keeping operating costs low. Amongst the wide range of equipment available for it are sand-channels, stowage panniers, general-purpose machine gun mount, tilt, ringmount/ rollcage, MK19 grenade launcher, GIAT 20mm gun, 0.50 Browning machine gun, 7.62mm machine guns, MILAN anti-tank missile, and communication and navigation equipment. Manufactured by Longline Management Services Ltd – who were later taken over by Ricardo Special Vehicles Ltd – the MRCV provides seating for four crew and can be fitted with under-vehicle protection against mines. It is air-portable/air-droppable in every body configuration. One Ten and 130-based MRCVs are also available.
(290: Land Rover Ltd, 1993)

290

291

292

Anti-tank vehicles based on the Series III Land Rover 88in 0.5 Ton Lightweight – one of an order sold to the Sudan government – and the Land Rover Ninety, both known as the Gunship. The special bodywork and fittings for these anti-tank vehicles is carried out by Marshall of Cambridge (Engineering) Ltd, and come standard with a 106mm recoilless light anti-tank gun M40A1 and mount M79, seats for the gun-crew, storage for both 106mm and 7.62mm ammunition, three blast-shields – to protect bonnet and wings against muzzle blast effects at low barrel depression – a dashboard-mounted barrel-clamp, split windscreen, barrel cleaning rods, pick and shovel. The right door of the Ninety Gunship is replaced by a fixed plate on which the spare wheel is mounted. The light bar just behind the front seats is fitted to restrict the amount of barrel depression when firing ahead.
(291, 292: Marshall of Cambridge <Engineering> Ltd, 1978, 1993)

293

Series III Land Rover 109in Soft Top with mounted TOW anti-tank weapon system, which is in use by the new Air Mobile Brigade from the Royal Dutch Army. Since the early Sixties the Land Rover has been used as an ideal weapons platform by many armed forces for a variety of anti-tank weapon systems such as the Viligant anti-tank missile, Swingfire anti-tank missile, 106mm light anti-tank gun, 120mm Wombat recoilless anti-tank gun and the modern range of HOT, MILAN, TOW and Carl Gustaf anti-tank missiles, due to its high cross-country capability and air-portability.
(293: Materieelist – Personnel Magazine of the Directorate of Material Royal Netherlands Army, 1993)

During the Sixties some track-shod Land Rover conversions were engineered and manufactured on request for the British MOD by J A Cuthbertson Ltd. They were intended for bomb disposal work, but later were also used for non-military purposes. This conversion consisted of a full-length heavy-duty box-section subframe on which the Land Rover chassisframe was mounted. Sprocket wheels were attached to the Land Rover's original axles instead of the normal road wheels. The subframe was fitted with fixed bogie wheels at the rear and steered bogie wheels at the front, which were positioned under the sprocket wheels. Thus four separate pairs of sprocket wheels and bogie wheels were formed, around each being fitted one driven track, of 12in width and consisting of 40 steel trackshoes, riveted on 6-ply nylon/cotton reinforced rubber belts. Because of the two heavily-steered front wheel couples, the vehicle was fitted with hydraulic power-assisted steering. A special trailer was also available on which the original Land Rover road wheels were fitted. This conversion kit was suitable for fitment to all Series II and IIA Land Rover 88in and 109in models, giving excellent stability and a smooth ride over the roughest terrain, with a maximum fording depth of 80cm. Because of the heavy weight of this conversion kit, neither the centre of gravity nor the cross-country stability changed much with regard to the original Land Rover, although the conversion was much higher. The unladen weight of an 88in Land Rover fitted with this conversion kit is 2,500kg.
(294: James A Cuthbertson Ltd, 1965; 295, 296: Nick Dimbleby, 1993)

294

295

296

297

298

299

300

A Land Rover half-track conversion manufactured in the early Eighties by Laird (Anglesey) Ltd for the British Army and known as the Centaur. In fact, this conversion was a combination of the front section of the Stage I Land Rover 109in V8 (including the V8 petrol-engine and original gearbox and transfer box with centre-differential) coupled with two tracks, two sprocket wheels and eight idler wheels derived from the British Scorpion range of light tanks manufactured by Alvis Ltd. Both front wheels and tracks were driven, which endowed a Land Rover with the best cross-country capability ever, combined with superb on-road steering quality at high speeds. This vehicle had a payload of 3 tons and was used as an ambulance, a command vehicle, a mine layer, a missile carrier, an artillery gun tractor and a 20mm anti-aircraft gun carrier. Two such conversions were sold to the Oman government and one was used for cold-weather trials in Norway. The rear section was 40cm wider than the original Land Rover front section. This heavy and wide track system, and the absence of armour and a turret, gave the vehicle an extremely low centre of gravity, which resulted in a superb admissible traverse gradient. However, the development of this conversion was abandoned in 1985 because the research and development costs were becoming high against an increasingly uncertain return. It was always hoped to put some into service with the British Army, but changing requirements and the development of the all-terrain vehicles made this unlikely, and only eight prototypes were built for rigorous trial purposes. Considerable interest was generated in a number of overseas markets, but without the UK in-service status it was unlikely to sell in any viable quantity, so the project was abandoned.
(297, 298: The Tank Museum, Wareham, 1983; 299, 300: Land Rover Ltd, 1983)

01

302

Royal Marine Land Rovers are specially prepared to bridge the distance between the landing craft and the beach during seaborne operations by deepwading. Vehicle preparation for such amphibious work entails protection against the salt water, thus necessitating special anti-corrosion treatment. Yet even the best treatment cannot protect against the full effect of the salt, which results in a very reduced life for the vehicles concerned. Further adaptations are required for the engine fan and air intake, fuel tank, gearbox, transfer box, brake fluid reservoir, battery vents and, of course, the electrical system. All these modifications are made by the Royal Marines themselves. Amphibious troops of many NATO-countries are trained at the Devon-based Amphibious Trials and Training Unit of the Royal Marines (ATTURM), using a fresh-water dip tank – or the sea itself – for deepwading experience.
(301, 302, 303, 304, 305: G Hydes, ATTURM, 1990)

303

)4

)5

306

During the mid-Fifties the British Army was looking at ways to improve the Land Rover's capabilities. Two systems were tried out to carry the vehicles across water with the aid of a lightweight flotation kit, which was transported in the back of the vehicles. The first system used an inflatable pontoon manufactured by RFD Ltd. Here the Land Rover was held in position on the inflatable pontoon by means of lightweight steel channels, which ensured an even spread of the load. These ramps also facilitated the exit and entry of the vehicle, this being achieved by deflating the rear compartment of the pontoon. The pontoon was used as a shuttle to cross wide and deep rivers, the crossing being effected by a rope being pulled on each side.
(306, 307: RFD Ltd, 1957)

307

308

A special air-portable Land Rover 109in 1 ton Scheme A fitted with the second system, which was also manufactured by RFD Ltd. This consisted of four airbags, which were fitted to pick-up points on the vehicle and inflated by its exhaust system. A steel propeller was mounted on the rear end of the vehicle's original propeller shaft – still driving the rear axle – to provide water propulsion.
(308, 309: RFD Ltd, 1959)

3

A Series II Land Rover 88in 0.5 ton – unladen weight 1,366kg – of the Royal Dutch Marines being airlifted by a British Navy Westland Wessex helicopter. To further improve the Land Rover's already high manoeuvrability, this vehicle is often airlifted by helicopters of many types, its air-portability being of particular value during rapid front-line operations.
(310: Hoofdkwartier Korps Mariniers, 1978)

310

During the late Eighties, with the support of Land Rover Ltd of the UK, Land Rover Australia – a division of Jaguar Rover Australia Ltd – developed a multi-purpose heavy-duty Land Rover 6x6, based on the One Ten, to meet the requirements of the Australian Army for a 2 tonne payload vehicle with good cross-country capability. In contrast with all other Land Rovers, this vehicle is fitted with coil springs at the front and semi-elliptic leaf springs to the rear and has a heavy-duty hot-dip galvanized chassisframe. For on-road operation, only the front and centre axle are permanently driven via the centre differential. For off-road operation the centre differential is locked and the third – rear – axle is engaged to give six-wheel drive. This conversion is suitable for many military roles, including that of cargo/personnel carrier, Rapier air defence system vehicle, ambulance, mobile workshop, long-range patrol vehicle, mobile command post, recovery vehicle, fire engine and gun tractor. It is also used in civilian roles as a light cement mixer, dump truck and fire engine. The weight of the basic vehicle (Chassis Cab) is 2,600kg; its total payload is 3,000kg inclusive of the weight of the rear body, load, crew/equipment and optional fittings. The maximum towing weight is 1,500kg, which makes a maximum combination weight of 7,100kg!
(311, 312, 313, 314: Land Rover Australia, 1993)

311

12

313

4

A launch vehicle conversion based on the Land Rover One Ten Pick Up. This vehicle is used by the armed forces to launch remote-controlled (pilotless) target aircraft to train the crew of Short Air Defence Units. The special launching equipment is installed in the Land Rover body by Marshall of Cambridge (Engineering) Ltd. The elevation of the launching-rail can be adjusted mechanically by a hand-operated vertical screw-spindle.
(315: Marshall of Cambridge <Engineering> Ltd, 1985)

315

A military Land Rover 109in staff vehicle capable of seating eight officers, based on one of the latest Series IIA Land Rovers with old-style grille and new-style headlight panels – produced by Pilcher-Greene Ltd in 1971.
(316: Pilcher-Greene Ltd, 1971)

316

317

A Land Rover 109in Forward Control fitted with two pto-driven drums between the front and rear road wheels, which could be lowered hydraulically when the vehicle became bogged down in the mud, to provide worm-drive. The British Army developed many interesting modifications for Land Rovers during the Sixties, but only prototypes were built of this novel drive system, for experimental and test purposes.
(317: The Tank Museum, Wareham, 1966)

Due to British Army needs in the late Sixties, a single-axled powered trailer was developed and manufactured by Scottorn Trailers Ltd, intended to be used exclusively with the military Series III Land Rover 101in Forward Control 1 Ton. The combination of this powered trailer with the 101in Forward Control appears in six-wheel-drive form, resulting in a very high cross-country capability. When loaded, this rear pto-driven trailer has enough power to push the Land Rover, even when the vehicle's propeller shafts are removed. It is attached to the Land Rover by the standard NATO hook and eye, and its propeller shaft is connected to the vehicle's rear power take-off via a quick detachable coupling positioned directly under the towing hook. The drive of the trailer is engaged and disengaged from the cab of the Land Rover, giving a choice of six or four-wheel drive, and its axle, differential, springs, shock absorbers, rims and tyres are the same as those fitted to the Land Rover. It is also equipped with an over-run brake, upon which a lockout device is used when the trailer drive (maximum articulation of 60deg in any direction) is engaged. Manufactured in large numbers, this trailer is available in 1 ton and 1.5 ton payload capacities. If 7.50 x 16in tyres are fitted, it is possible to use it in combination with the normal-control Land Rover. During the early Seventies, a number of similar trailers were manufactured by Rubery Owen and Company Ltd using a different articulated quick-release attachment.
(318, 319, 320: Reynolds Boughton Ltd, 1978)

318

319

320

321

In 1988 the Hymatic Engineering Company Ltd were involved in converting two military Land Rover Ninetys to provide a high-pressure thermal imager gas-bottle-charging facility. Most thermal images used by the British and other NATO forces require a supply of very high-pressure gas (3,500psi) for their operation and, in order to support thermal images in the field, a highly flexible and responsive vehicle was equipped with Hymatic-designed compressors and special gas filtration. Only two prototypes were built, one of which was sent to Afghanistan, the other vehicle remaining in the possession of Land Rover Ltd. Options to purchase larger quantities of these specially equipped vehicles were never taken up by the British Army.
(321, 322, 323: The Hymatic Engineering Company Ltd, 1988)

322

32

Shown at the British Army Equipment Exhibition (BAEE) of 1993, a Defender One Ten Soft Top, with raised and widened rear section, built according to military specifications in order to enlarge the load capacity.
(324: Nick Dimbleby, 1993)

324

325

Land Rover 130 Chassis Cab-based troop carrier supplied to the government of Rwanda. The wide dropside body with high tilt is manufactured and fitted to the chassis by Marshall of Cambridge (Engineering) Ltd.
(325, 326: Marshall of Cambridge <Engineering> Ltd, 1990)

326

Municipal use...

327

Refuse disposal unit conversion for which both Series IIA and Series III Chassis Cab 109in Land Rovers are used as a basic vehicle, manufactured by The Eagle Engineering Company Ltd. The welded steel tipping rear body, with a capacity of 2.25cu m, has two sliding shutters on each side and a large, lockable, top-hinged tailgate at the rear which allows for easy discharge when tipping. This unit was fitted with hand-operated tipping gear, including a long vertical screw-spindle for lifting the front end of the tipping body. This conversion was also available as a normal open tipper unit without the sliding shutters, which was fitted with a top-hinged low tailboard.
(327: Dennis Eagle Ltd, 1972)

328

Single-axle refuse disposal trailer, of 2.25cu m capacity, fitted with a hand-operated hydraulic tipping system, two sliding shutters on each side and a full-size top-hinged rear tailgate. Like the trailer opposite (329), this unit was designed and manufactured by Scottorn Trailers Ltd.
(328: Reynolds Boughton Ltd, 1978)

329

Twin-axled tanker trailer designed for the removal of liquid waste and sludge. It is fitted with a full-size hinged rear door and a pressure/vacuum pump, which is attached to the front end of the tank. The tank itself has a capacity of 1,350 litres.
(329: Reynolds Boughton Ltd, 1978)

Tipper conversion, engineered and manufactured by Special Vehicles, for which the Defender 130 Truck Cab and Crew Cab chassis are used. The flat-bed tipping rear body of the Truck Cab version is 1.90m wide and 2.70m long and has a payload capacity of 1.25 tons, the Crew Cab version necessarily having a reduced the rear body length and payload capacity. The tipping rear body has a rigid steel frame fitted with dropsides, to which are attached heavy-duty latches and a top-hinged tailgate, both made of extruded aluminium, which can be raised by optional weld-mesh panels to give an enlarged carrying capacity.
(330: P D Stevens & Sons Ltd 1993; 331: Nick Dimbleby, 1993)

330

331

The electro-hydraulic tipping system is engineered by Smiths Industries Hydraulics Company Ltd of Witney, which allows tipping in a rearwards direction only, controls for which are positioned in the cab. This conversion is fitted with a heavy-duty suspension and is also available with a non-tipping dropside body, as shown in (97) and (96). Volume production of all these three conversions is by P D Stevens & Sons Ltd.
(332: R de Roos, 1993)

332

Police patrols...

333

A Land Rover Discovery converted for the Algemene Verkeers Dienst (AVD) – subsidiary of the Dutch police – for general road traffic services such as rescue work and technical traffic inspections. The equipment for this vehicle includes anchors, an electrical winch with steel cable, 12kg fire extinguishers, firemen's clothing and helmets, a high-lift jack, towing cable – strong enough to tow lorries – 2.5-ton hydraulic jack, small and large first-aid kits, a comprehensive tool set, 20 orange road cones, a rotating beacon, crowbar, torches, aluminium blankets and a tool to measure the height of loaded lorries to check that the maximum permitted height is not exceeded.
(333: R de Roos, 1991)

334

A less military-looking version of the Penman Engineering Hussar armoured vehicle range. Known as the Polisec, this vehicle is a 6x6-driven armoured personnel carrier based on the One Ten. It provides the same full armoured protection as the Hussar APC, mentioned earlier, for an eight or 12-man police team.
(334: Penman Engineering Ltd, 1990)

335

One Ten-based fully armoured police vehicle with 5-9mm special armoured steel monocoque hull and 35mm laminated bulletproof glass windows, both providing protection against 7.62mm NATO ball ammunition, blast and grenade fragments. Extra protection is given to the fuel tank by a self-repairing system which seals small bullet holes automatically, and to the crew in the front by an additional armoured belly plate. The run-flat capability of the road wheels provides a minimum loss of steering control after being hit by rifle fire. Manufactured by Glover Webb Ltd, this vehicle can be fitted with a comprehensive range of optional equipment suitable for other applications such as airfield security, anti-terrorism, military and paramilitary duties.
(335: Nick Dimbleby, 1992)

336

Other Land Rover conversions by Glover Webb Ltd based on the One Ten related to police/anti-terrorist and riot control work include rebel trooper vehicles...
(336: Glover Webb Ltd, 1992)

...quick building access/assault vehicles...
(337: Glover Webb Ltd, 1992)

337

...armoured anti-terrorist vehicles...
(338: Glover Webb Ltd, 1992)

338

...and prison vans.
(339: Glover Webb Ltd, 1992)

339

340

341

Police vans based on the Series II Land Rover 88in chassis and used from 1958–66 by the police force of the Mersey Tunnel Authority to retrieve broken-down vehicles. The colour scheme of these vans was cream with red front wings and roof; police vans built from 1956–58 were based on the Series I Land Rover 88in chassis and were painted in red. The special bodywork for both Series I and Series II-based vans was carried out by Noel Lacey Ltd. At the time of writing the Mersey Tunnel Authority has white Defender Ninety Tdi Hard Top tunnel police vehicles in service, with red and blue striping.
(340, 341: Merseyside Passenger Transport Authority, Mersey Tunnels, 1961)

342

Dinky Toys model of the red Series I Land Rover 88in-based Mersey Tunnel police van, produced in large numbers in Liverpool by Meccano Ltd during the Fifties and Sixties.
(342: R de Roos, 1993)

343

In 1993 Zumro (UK) Ltd fitted two Land Rover Discoverys with a roof-mounted Stem-Lite mast, then shipped them out as part of an order for the Moscow traffic police. They have since taken orders for several identical vehicles. The Stem-Lite system consists of a mast with beacon and floodlights, an auxiliary alternator, a control panel, additional hand-throttle cables, a voltage regulator and a volt meter.
(343, 344: Zumro <UK> Ltd, 1994)

344

Rescue and emergencies...

346

345 During the late Sixties Merryweather & Sons Ltd manufactured rescue units based on the Series IIB Land Rover 110in Forward Control Chassis Cab which were fitted with 2.6-litre, six-cylinder Rover petrol engines. A unit's GVW was 4,200kg and its fuel consumption approximately 9mpg! They were in service by fire brigades mainly for motorway use, and had a crew of four including the driver. Each had a fully closed, red/silver-painted body fitted with four doors and seven light-alloy roller shutters and came complete with this
347 impressive list of tools and equipment:

EXTERIOR
Searchlight
Blue rotary flashing light

DRIVING CAB
Portable spotlamp
Wireless set
3 appliance logbooks

CREW CAB
2 Forster hand lamps
Railway signalling horn
50ft GP line
100ft GP line
120ft lowering line
2 3lb BCF extinguishers
4 Surcoats
2 3lb BCF extinguisher refills

LOCKER NO 1 TRANSVERSE
2 first-aid haversacks
Resuscitator
2 CA sets
2 tins anti-dim
Small crowbar
Large crowbar
7lb sledge hammer
Pick helve
Pick head
Metal cutting axe
Eclipse saw
Spade
Shovel
Bolt cropper
Wire cutter
Grapnel
Transhailer
Asbestos blanket
Quick-release knife
Bosuns chair and sling
3 15ft air lines
Vehicle starting handle
Air reel 2 CA cylinders
Impactor air cutter
Cengar air saw
12-ton jack handle

LOCKER NO 3 REAR
4 Mitralux searchlight legs
2 cluster lights
2 spare bulbs 110 volts
2 spare Mitralux bulbs
2 ball and socket lamp holders
6 cable hooks in box
1 Nife searchlight
2 tripods
2 searchlights
2 reels with 12ft cables
Two-way junction box cable + reel
Mitralux portable generator
2 150in cable and reel

LOCKER NO 4 CENTRAL TUNNELS
Short extension ladder
Folding stretcher
GQ Para stretcher and slings
2 steel shod levers
2 wool blankets

Foil blanket
Plastic body sheet
Wheels for compressor
Milamatic air compressor
Inner sliding shelf
Hearth kit containing:
Bolster chisel
1in cold chisel
½in cold chisel
Hacksaw
12 blades
4lb mason hammer
Cockerel saw
Insulated pliers
8in screwdriver

GLOVEBOX
2 pairs PVC gloves
2 pairs industrial gloves
4 eye shields

ELECTRICAL KIT
Pair rubber gloves
12in screwdriver
8in screwdriver
Insulated pliers
Phillips screwdriver

BOX ENGINEER'S TOOLS
Ball peine hammer
Pliers
Large screwdriver
Small screwdriver
Tommy bar
3 box spanners
Adjustable spanners
5 DE spanner
SE spanner
Pad saw
14in Stillson
Tinsnips

LOCKER NO 2 BOTTOM
1½gall can petr/oil mix
15ctw trolley jack + handle

Flexiforce hydraulic lifting kit

LOCKER NO 5 REAR
20in anchor cable
2 blue flashing Tildawn lamps
2 police accident signs
Wire sling double leg
2 ground anchors
2 tripods
40ft wire cable on reel
4 3-ton D-shackles
6 large steel spikes
4 small steel spikes
Tirfor (T35) and handle
Snatch block
Bow shackle

LOCKER NO 6 BOTTOM
4 wooden blocks
2 wheel chocks
12-ton jack

A pto-driven Plummett winch was also fitted, with 230m cable, but a first-aid water tank, pto-driven water pump, suction and delivery hoses were omitted. However, Merryweather & Sons Ltd offered a fire engine version based on the same vehicle, but with a slightly changed body configuration. This Fire Warden had a minimum turning circle of 14.6m, and was fitted with a midships-mounted Merryweather MB1-4 waterpump, of 600gall/min maximum output, with automatic water-ring priming. It carried two 65-gallon water tanks and an impressive set of tools and equipment; a 5m light-alloy ladder came as standard. The body – available with a two-man cab or four-man crew cab – contained the suction inlet, delivery outlets, hose reel and controls at the rear. The rear bodywork could be lifted completely clear of the chassis so that it was easily accessible when repair work had to be carried out. HCB Angus Ltd also made fire engines which were based on the Series IIB Land Rover 110in Forward Control at this time.
(345, 346, 347: Marcus Kelly, 1972)

348 Series III Land Rover 109in Station Wagon-based road accident vehicle, manufactured by Angloco Ltd, used by several fire brigades and equipped to provide assistance at the scene of serious road traffic accidents.
(348: Angloco Ltd, 1980)

349

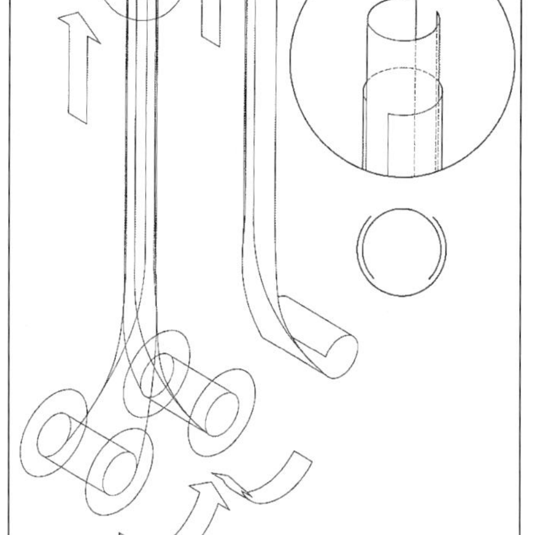

350

Six-wheel-drive, high-roof road traffic accident rescue vehicle, manufactured by Reynolds Boughton Ltd. This was supplied to the Dorset Fire Brigade in 1987 to provide rescue assistance for road, railway and air accidents, and is equipped with the following: electrical front-mounted winch, separate halogen floodlights on tripods, air-cushion jacks, a roof-mounted Stem-Lite for a 2.5m elevation of the halogen floodlights and a blue xenon flashlight – which provides an earlier warning to approaching traffic – two big roof-mounted blue xenon flashlights, large roller shutters left and right and fire extinguishers. The crew cab provides a seating capacity for four including the driver. The Stem-Lite, which is mounted on many Land Rover, Range Rover and Discovery-based rescue, police, fire-fighting and emergency vehicles, is a unique mast system which is *not* telescopic. It consists of two resilient spring strips of stainless steel which form a tubular shape of high strength with a diameter of 2in (50mm). During transit these strips are flat and stored on two electromotor-driven storage drums (350). This STEM (Storable Tubular Extendible Member) system was developed by the Canadian company Spar Aerospace Products Ltd, in 1960, and used initially as a communication antenna for the Gemini spacecraft. Several models of the Stem-Lite – a product of the Dutch company Zumro bv – are available today, fitted with powerful floodlights and/or rotating beacons. In the transit position the Stem-Lite mast has an exterior height of 45cm and diameter of 38cm.
(349: Nick Dimbleby, 1993; 350: R de Roos, 1993)

351

352

Another type of six-wheel-drive Land Rover rescue vehicle, with a telescopic floodlight mast and light-alloy extension ladder mounted on the unraised roof, also manufactured by Reynolds Boughton Ltd.
(351, 352: Reynolds Boughton Ltd, 1989)

Road maintenance...

High-capacity snow-clearing equipment such as snowblowers, which could be mounted to the front of the Land Rover, were manufactured in the Fifties by British Rotary Snowplough Ltd. A separate engine was installed in this type of snow-blower, and was not powered by the power take-off. In the Sixties, Rolba Ltd produced hydraulically pto-driven snowblowers which could be used with the Land Rover. (353: National Motor Museum, Beaulieu, 1958)

353

354

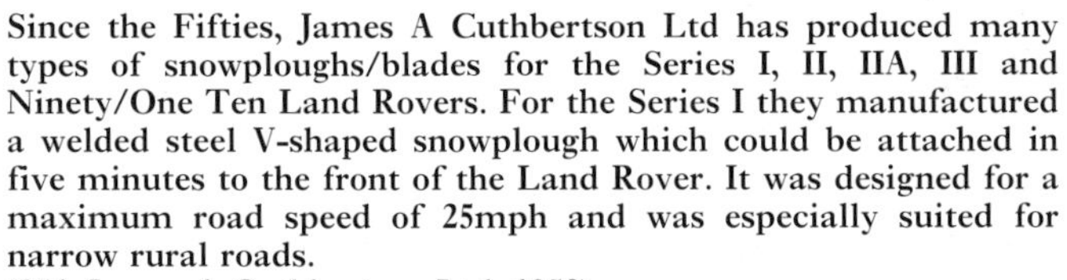
Since the Fifties, James A Cuthbertson Ltd has produced many types of snowploughs/blades for the Series I, II, IIA, III and Ninety/One Ten Land Rovers. For the Series I they manufactured a welded steel V-shaped snowplough which could be attached in five minutes to the front of the Land Rover. It was designed for a maximum road speed of 25mph and was especially suited for narrow rural roads.
(354: James A Cuthbertson Ltd, 1953)

355

A Cuthbertson reversible scraper blade, designed for fast and efficient snow clearance and intended for wider roads. The angle of the blade is adjustable to the left and right, and the scraper blades are available with cab-operated hydraulics.
(355: James A Cuthbertson Ltd, 1964)

356

357

358

Three views of 2-metre wide electro-hydraulically operated snowploughs, which fit all types of Land Rovers and Range Rovers without drilling or making alterations to the vehicle and are manufactured by Econ Atkinson Ltd, who have been producing winter maintenance equipment since 1950. Angles up to 30deg to the left or right are achieved by means of a quadrant-frame, which remains at both ends of the blade in all positions. A self-levelling device enables the blade to operate while following road contours, and when raised it is carried parallel to the road.
(356, 357, 358: Econ Group Ltd, 1967, 1979, 1986)

359

An Econ Atkinson trailed gritter attached to a Series III British Army Land Rover 109in. The conveyor and distributor are hydraulically powered by the trailer's axle-driven hydraulic pump or by the pump in the Land Rover's own hydraulic system. Its carrying capacity is 2 tonnes and its spread width is adjustable through the 0–11m range.
(359: Econ Group Ltd, 1979)

Reinforced polyester scraper blade for snow clearance work, designed to fit the Land Rover One Ten and manufactured by the Dutch company Nido Universal Machines bv. The blade's specially-shaped surface, combined with a high road speed and an angled mounting to the vehicle, ensures that all snow is cleared to the side of the road without drift-snow falling on the vehicle's bonnet. The mounting consists of a large horizontally-sprung hinge which can absorb the impact of hard, fixed road obstacles without damaging the scraper blade: when the blade hits an obstacle it rolls forward, then is quickly repositioned by the spring. The weight of the complete system for a 2.1m work width is 265kg and a heavy-duty front suspension is required. The scraper blade's ground pressure is adjustable.
(360, 361: Nido Universal Machines bv, 1993)

360

361

A wide range of hydraulically-powered gritters and trailer gritters attachable to Land Rovers are manufactured by Nido Universal Machines bv, with an adjustable spread width in the 2 to 6m range and a carrying capacity of 1 tonne. Their on-road spread density is adjustable from 40 to 200g per sq m.
(362: Nido Universal Machines bv, 1993; 363: R de Roos, 1993)

362

363

364

In the early Sixties, Heenan & Froude Ltd, of Worcester, developed two systems which could be fitted to the Land Rover to create a relatively cheap and light railway-engine for shunting work on factory-site railway tracks. Then it could be used as a service unit which came equipped with a hydraulic work platform, welder equipment, air compressors or tool set for railway maintenance. The first system consisted of four large flanged rail wheels – instead of normal road wheels – which were fitted directly to the wheel hubs of the Land Rover.
(364: James L Taylor Collection, 1961)

365

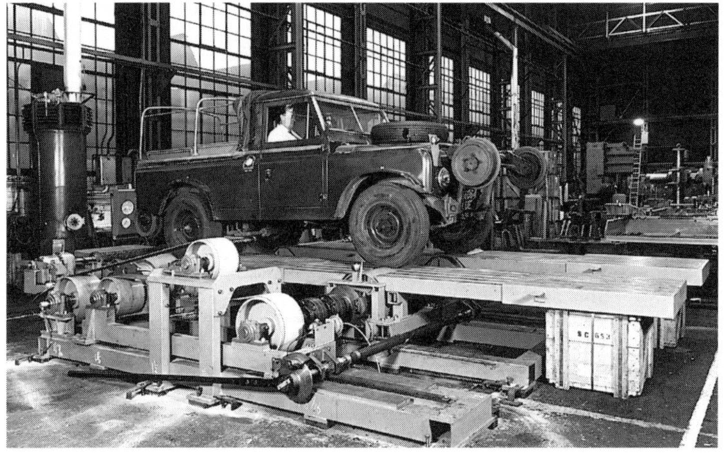

366

The second – a road/rail system – had a front and rear-mounted rail guidance system which could be lowered and raised hydraulically. Traction was provided by the four widened road wheels. Three of these conversions were made, of which at least one was exported to South Africa. Here it is shown mounted on a Series IIA Land Rover 109in Pick Up, which is positioned on a chassis dynamometer, which was also produced by Heenan & Froude Ltd and supplied to the Rover Works in 1962.
(365, 366: Froude Consine, 1962)

367

368

A road/rail guidance system which could be fitted to the Land Rover Ninety/One Ten, Discovery and Range Rover models was also developed and manufactured by Wickham Rail Ltd. This was mounted to emergency vehicles, hospitalization units, fire engines, police and railway maintenance vehicles to provide quick access to railway locations where there were no normal roads. To fit this system to the vehicle's chassisframe it was only necessary to remove the front bumper, the front system then being bolted onto the existing box section's sidemembers. The rear system was bolted directly to the chassisframe's rear crossmember. The double bogey wheels ensured a good ride and excellent performance up to a speed of 30mph even on tight curves and poor-quality third-world railways, although there were no springs fitted in the system. Lowering the bogey wheels on a crossing made transition from road to rail simple. Traction was still provided by the normal 4x4-driven road wheels. This system was also used by the Sales Department of Land Rover Ltd to demonstrate to the press the enormous torque of the new 2.5-litre turbo diesel engine fitted in a Discovery, pulling rail coaches with a total weight of 170 tons. For normal road use, it was unnecessary to remove the road/rail guidance system: raising the bogey wheels alone was sufficient.
(367, 368: Wickham Rail Ltd, 1992)

Trailers...

369

A variety of manufacturers, including Scottorn Trailers Ltd, Dixon Bate Ltd, GKN Sankey Ltd, Pressed Steel Company Ltd, Rubery Owen Company Ltd, York Trailers Ltd, Ifor Williams Trailers Ltd and King Trailers Ltd, have developed and produced civil and military trailers exclusively for use with the Land Rover. However, the first specially designed trailer for this vehicle was manufactured by J Brockhouse and Company Ltd in the early Fifties for the Series I. This single-axled trailer for farmers had a steel chassis with a fully welded body of pressed-steel panels, a payload capacity of 750kg. Illustrated here is a custom-built general purpose cargo trailer with a payload capacity of 1.5 tonne fitted with a standard NATO coupling and canvas cover.
(369: Reynolds Boughton Ltd, 1978)

370

A general-purpose cargo trailer with a payload capacity of 2 tonnes. Like the trailer in the previous illustration it was designed and manufactured by Scottorn Trailers Ltd, a division of Reynolds Boughton, who supplied a varied range of standard trailers and could adapt or design units to suit specific customer requirements. The trailers were subjected to continuous test programmes both in the UK and elsewhere, particularly in Africa and Asia, where much experience was gained by their operation with military forces.
(370: Reynolds Boughton Ltd, 1978)

371

372

373

Ground-level loading trailer manufactured by King Trailers Ltd and available in capacities from 0.75 to 6 tonnes. A hydraulically-operated suspension allows the trailer to be lowered to the ground, powered by the hydraulic pump or the vehicle's hydraulic system. When fitted with a winch this trailer is ideal for recovering heavy non-powered vehicles and equipment. Adjustable-height towbars are available to allow the trailer to be used by vehicles with different towing heights.
(371, 372, 373: King Trailers Ltd, 1992)

374

The adaptable lightweight tipper trailer from King Trailers Ltd, available from 1.5 to 3 tonnes capacity. The hydraulic ram for the tipping operation is powered hydraulically or electro-hydraulically by the towing vehicle; the rigid dropsided bodies are manufactured from either steel or timber.
(374: King Trailers Ltd, 1992)

375

A twin-axled tipper trailer, made by King Trailers Ltd, with a similar range of capacities and a loadbed 3m long by 1.65m wide, designed for forestry work.
(375: King Trailers Ltd, 1992)

376

A general-purpose trailer, for which a number of options were available, amongst them reinforced extended steel sides or drop-sides, detachable loading ramps and reinforced steel chequer deck plates.
(376: King Trailers Ltd, 1992)

377

378

Ifor Williams horsebox trailers from the Hunter range, for the transport of two horses, constructed totally from high-quality, durable materials. Fully galvanized steel is used for both the chassisframe and bodyframe. The roof, front and door panels are made from patterned aluminium, and resin-coated plywood is used for the body and ramp panels. These trailers came standard with a groom's door, front ramp with gas-strut assister, rear ramp with assister springs, doors above the rear and front ramps, breast bars and breeching bars, side padding, side kicking boards, removable rubber mats on the floor for easy cleaning, rubber matting on ramps, toughened glass windows, a GRP front roof cone, beam axles with maintenance-free heavy-duty leaf springs and an auto-reverse braking system. Ifor Williams Trailers Ltd, who also produce a highly varied range of agricultural and commercial trailers up to a gross weight of 3,500kg, have been manufacturing horsebox trailers for more than 30 years. Vehicle-mounted double horseboxes and pick-up canopies are also available. Land Rovers are ideal vehicles for such towing work because of their high-torque engine, permanent four-wheel drive and rigid chassisframe.
(377, 378: Ifor Williams Trailers Ltd, 1993)

Another example of the aforementioned Ifor Williams horsebox, but this time with a narrower trailer, designed for the transport of one horse.
(379: Ifor Williams Trailers Ltd, 1993)

379

One of 14 available types of livestock trailers fitted with aluminium ramp gates, double inspection doors, aluminium tread-plate floor covering, ramp springs, aluminium body panels; the chassis and bodyframe, ramp, inspection door and roof panels are all made of galvanized steel. A variety of optional equipment is available including sheep decks, cross divisions and centre partition/breast bar/breeching chains.
(380, 381: Ifor Williams Trailers Ltd, 1993)

380

381

382

383

384

A pair of twin-axled Ifor Williams trailers with tipping and low-loading bodywork, and one of the company's single-axled general-purpose trailers, all built to the same quality standards.
(382, 383, 384: Ifor Williams Trailers Ltd, 1993)

Vehicle recovery and service...

Since the Fifties many types of hand-operated vehicle recovery cranes have been manufactured by Brockhouse Harvey Frost Ltd to fit both short and long-wheelbase Pick Up Land Rovers. This recovery crane unit consisted of a heavy-duty steel tube frame, a hand-operated self-braking lift-gear unit, an anti-sway A-frame/distance A-frame, a sling with chains and hooks, a heavy-duty crane block for double cable runout and a 7.2m steel cable with hook. The maximum safe working load for this crane was 1.5 tons – strong enough to handle any car or light van – but if used with a short-wheelbase Land Rover a maximum of 400kg lift capacity was allowed, while with a long-wheelbase version, a maximum of 900kg lift capacity was allowed in order to avoid excessive stresses in the sidemembers of the vehicle's chassis-frame, and to retain sufficient on-road steering quality.
(385: FKI Bradbury Ltd, 1965)

385

Part of an order of 50 vehicle recovery crane conversions for a Libyan customer, based on Series IIA Land Rovers.
(386: FKI Bradbury Ltd, 1968)

386

A more recent electric-powered Land Rover recovery crane, manufactured by Harvey Frost Epco Ltd. This had a remote-control switch with a 3m cable which meant that the crane could be operated from the best position possible.
(387: FKI Bradbury Ltd, 1980)

387

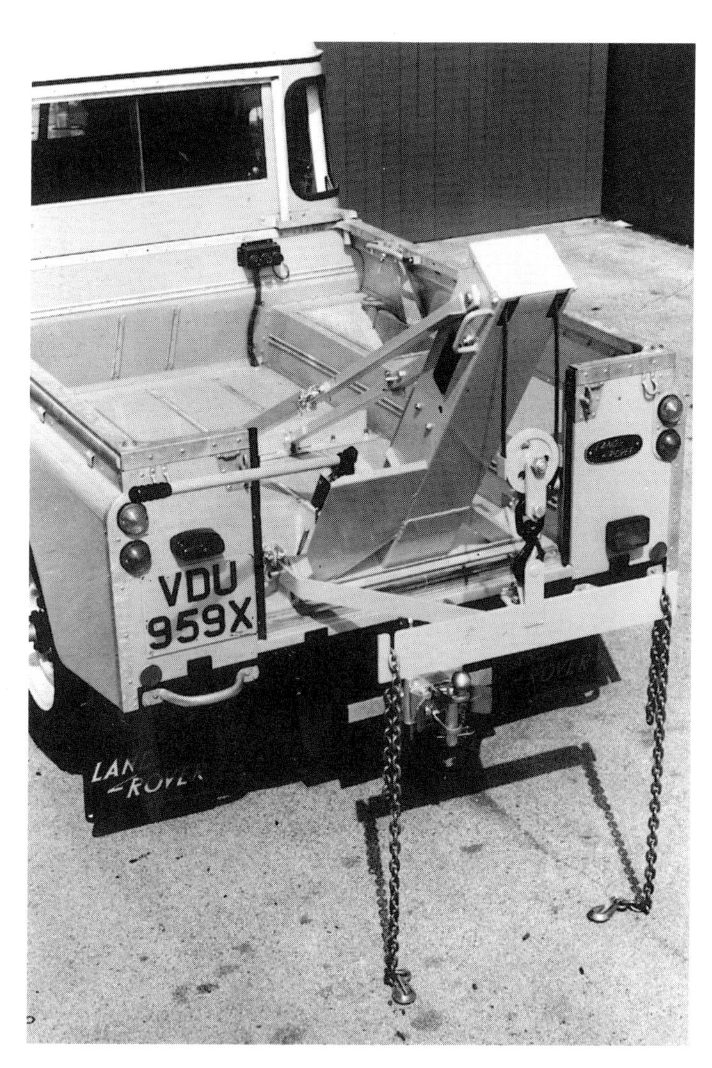

388

389

Hand-operated hydraulic recovery crane for both Series III 88in/109in and Ninety/One Ten Pick Up Land Rovers, manufactured by Dixon Bate Ltd. This 1.5 tonne recovery crane, known as the Tuckaway-R, is actuated by a hand-operated double-acting hydraulic pump. It can be quickly folded to lie flush on the vehicle's floor when not in use. An automatic lock which takes the load off the hydraulic system when towing is fitted. This vehicle recovery crane is supplied with lifting hooks, chains, A-frame and spreader bar. Rear suspension assisters are also required. It is used by the military, police and many local garages with recovery services. Articulated single-axled trailers with a high swan-neck attachment – which could only be used with Land Rovers – were manufactured by the same company in the Sixties, who for many years also produced a comprehensive range of towing couplings, jaws and pintles for both civilian and military Land Rovers.
(388, 389, 390, 391: Dixon Bate Ltd, 1992)

390

391

392

Hydraulically-operated vehicle recovery system for long-wheelbase Land Rovers, manufactured by Bristol Metal Contracts Ltd – Brimec (UK) Ltd – and known as the Towlift. This well-designed unit, with a payload of 800kg, is claimed not to damage spoilers, plastic bumpers and suspension parts of lifted vehicles and consists of a hydraulically-lifted and extended beam on which a wheelframe is mounted. This has flattened steel tubes at the rear, allowing it to be easily pushed under the stricken vehicle's wheels by extending the beam. Once the wheels are in the frame, it is lifted 50cm from the ground. The wheels are then strapped securely to the frame by a high-tension strap system. At this point the beam is retracted to its original length to prepare the vehicle for transport. If necessary, a standard fitted winch with a low horizontal steel cable is available to pull a vehicle into the wheelframe. When the beam and frame are not in use, they are hydraulically folded back into a vertical position. Land Rovers with the Towlift system fitted are equipped with heavy-duty rear suspension. A fleet of 12 One Ten-based Towlift vehicle recovery units were supplied to the Hong Kong authorities to maintain the flow of traffic in the new Hong Kong-Kowloon tunnel.
(392, 393, 394, 395: Brimec <UK> Ltd, 1992)

393

394

395

A heavy-duty 6x6 Crew Cab recovery crane conversion based on the Defender One Ten. Its equipment is manufactured by Wreckers International Ltd and fitted by Special Vehicles. A Ramsay 8,000lb electric winch is the principal component; a second, Husky 8,500lb electric drum winch, manufactured by Superwinch Ltd (Fairey), is fitted to the front.
(396: Land Rover Ltd, 1993)

396

← 397

398

Maintenance and service is very important for expensive and heavy equipment used in the construction, mining, forestry and civil engineering fields, such as excavators, earth-moving machinery, bulldozers, diggers, scrapers and dumper trucks, especially in remote and demanding terrain. To give all necessary maintenance and service to this equipment on-site, Tecalemit Garage Equipment Company Ltd designed and manufactured a mobile lubrication service unit conversion based on long-wheelbase and 6x6 Land Rovers. These conversions can be fitted to customer requirements with a two-stage petrol-driven 10 bar air compressor, air-powered grease and oil pumps, tyre inflator, oil sprayer, high-pressure wash pump, air, oil and grease hose on hose reels, compressed-air tools, oil and grease drums and battery charger. Optional extras include a canvas tilt for all-weather and dust protection, and a lifting gantry complete with a pulley block and lifting slings.
(397, 398, 399: Tecalemit Garage Equipment Co Ltd, 1980)

399

400

A mobile service unit conversion based on a 6x6-driven One Ten chassis used by the British Army and known as the Hotspur Trekker 6x6. Its chassis conversion is carried out by Hotspur Armoured Products Ltd (Penman Engineering Ltd).
(400, 401: Tecalemit Garage Equipment Co Ltd, 1986)

401

402

A Series III Land Rover 109in Chassis Cab mobile lubrication unit conversion by Reynolds Boughton Ltd. Mounted to the four or eight-cylinder-engined chassis, it is designed for the maintenance and servicing of a wide variety of vehicles, plant and machinery. Operators could forward servicing equipment directly to the site over varying conditions of terrain, both on and off road, with a minimum of delay. A similar type of unit providing workshop equipment for maintenance and repair was also available.
(402: Reynolds Boughton Ltd, 1979)

403

404

The Automobile Association has been using Land Rovers for almost as long as they have been made, their familiar yellow-painted vehicles carrying the bold AA motif being a source of reassurance to motorists in the UK for nearly half a century. Early examples were used in the main on urban service but in the mid-Sixties they also began to provide a motorway breakdown service, as demonstrated above by this Series II Land Rover 88in Hard Top patrol vehicle.
(403, 404: The Automobile Association, 1952, 1966)

405

In recent years AA Land Rovers have been an equally familiar sight in rural areas, as evidenced by this Ninety Hard Top.
(405: Nick Dimbleby, 1993)

406

In the late Seventies the trailer manufacturer Ibbett Engineering Ltd obtained some new 109in V8 Stage I Chassis Cabs and ingeniously they cut the chassisframe just behind the transfer box and added their own design of rear frame, loadbed and suspension to create a kneeling car transporter, the rear suspension being hydraulically adjustable in height to enable the loadbed to be lowered to the ground. An electrically driven winch was fitted to the front of the loadbed to pull the broken-down car on board before the suspension was raised again. As it was not possible to power the rear axle on these conversions, the Land Rover's central differential was locked to provide permanent front-wheel drive. Unfortunately, however, their front-axle load proved to be insufficient for optimum performance when transporting another vehicle, although the AA did use them for a period in the early Eighties.
(406: Richard de Roos, 1994)